D0855053

Environmental Sampling for UNKNOWNS

Environmental Sampling for UNKNOWNS

Kathleen Hess

LEWIS PUBLISHERS

Boca Raton New York London Tokyo

Acquiring Editor: Ken McCombs
Project Editor: Albert W. Starkweather, Jr.
Marketing Manager: Greg Daurelle
Cover Designer: Denise Craig
PrePress: Kevin Luong
Manufacturing Assistant: Sheri Schwartz

Library of Congress Cataloging-in-Publication Data

Hess, Kathleen
Environmental sampling for unknowns / Kathleen Hess.
p. cm.
Includes bibliographical references and index.
ISBN 1-56670-171-6 (alk. paper)
1. Environmental sampling I. Title.
RA566.H47 1996
615.9'07--dc20

96-13906
CIP

Direct all inquiries to CRC Press, Inc., 2000 Corporate Blvd., N.W., Boca Raton, Florida 33431.

International Standard Book Number 1-56670-171-6
Library of Congress Card Number 96-13906
Printed in the United States of America 1 2 3 4 5 6 7 8 9 0
Printed on acid-free paper

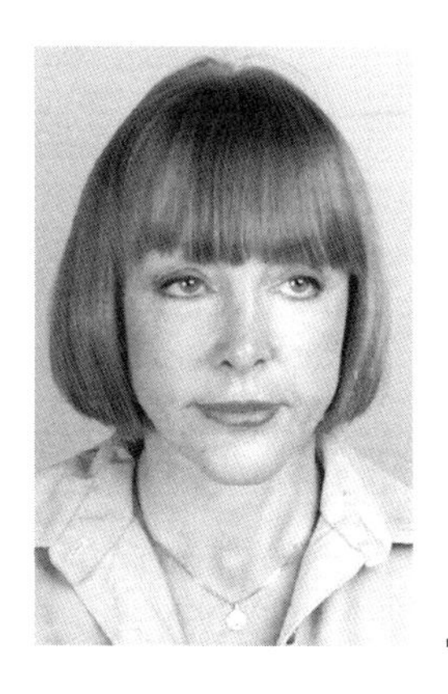

ABOUT THE AUTHOR

Kathleen Hess is president/owner of Omega Southwest Environmental Consulting. In 1972, she received her Bachelor of Science degree in Microbiology with a minor in chemistry from Oklahoma State University. After serving as an officer in the Air Force for three years, she returned to school and earned a Master of Science degree (1979) in Industrial Hygiene from the College of Engineering at Texas A&M University. Her research involved an animal toxicological study and was conducted at the College of Veterinary Sciences.

Ms. Hess worked as a consultant for Firemen's Fund Insurance Companies until 1984, shortly after passing the certification exam conducted by the American Board of Industrial Hygiene. During these five years, Ms. Hess had the opportunity to become involved in a variety of unique environmental problems, including indoor air quality concerns in an 800-occupant office building and information gathering for performing environmental site assessments and assessing waste. All this and much more carried over to her private consulting business.

Ms. Hess has since conducted numerous Phase I environmental site assessments and published a book concerning the topic. Ms. Hess has actively pursued obscure sources of information and training to better address the complex issues that face environmental issues, indoor air quality, and multiple chemical sensitivity. Her record to date has been that of identifying sources of indoor air quality problems (oftentimes other than air handling units) and has provided appropriate point source solutions to these very complicated problems. It took some time to get to this point, but some of the information which was collected and has been used successfully is presented within this book.

ACKNOWLEDGMENTS

There are many who have contributed to this publication and dedicated time to offer assistance in order to make this book the best it could be. Many are mentioned in the Reviewers' List, and some deserve special mention. One of these was Dr. Dick Thompson who patiently suffered the shock of learning the contents were not all chemistry. As a seasoned industrial hygienist, his comments were to the order of "very interesting" and "never have I read the likes of this book!" This was gracious comment and a warning to others.

Others who have been supportive of me in my frustrated efforts to pry out information from the various sources include my editor, Ken McCombs, his assistant, Susan Alfieri, and my family. The demon from within did strike a few times, and still respondents persisted to support me. To these and many others whom I may not have mentioned, I offer my gratitude and heart-felt thanks.

ACKNOWLEDGMENT OF TECHNICAL REVIEWERS

The Sea of Unknowns:
HazCat System
Robert Turkington
HazTech, Inc.
San Francisco, California

Allergenicity, Plant Pollen, and Mold Spores
Jim Thompson, Ph.D.,
Miles, Inc.,
Western Oregon State College
Monmouth, Oregon

Viable Microbial Allergens
George Morris, Ph.D.
Pathcon Laboratory
Norcross, Georgia

Airborne Pathogens/Microbial Toxins
George Morris, Ph.D.
Pathcon Laboratory
Norcross, Georgia

Animal Allergenic Dust
Thomas J. Lintner, Ph.D.
Vespar Laboratory,
subsidiary of ALK
Spring Mills, Pennsylvania

Rapid Microbe Fingerprinting:
Polymerase Chain Reaction
Claudia Thio
Perkin Elmer
Alameda, California

Microbial Identification System
Myron Sasser, Ph.D.
MIDI
Newark, Delaware

Volatile Organic Chemicals
Hernden Williams, Ph.D.
Radian Corporation
Austin, Texas

Eugene Kennedy, Ph.D.
Ardith A. Grote, Ph.D.
NIOSH Research Facility
Cincinnati, Ohio

Carcinogens/Mutagens
Richard H. C. San, Ph.D.
Microbiological Associates, Inc.
Rockville, Maryland

PREFACE

As the year 2000 approaches, advances in environmental sampling technology have been hard sought, generally remaining part of well-guarded, jealously-protected private libraries. Although some of these libraries are more complete than others, the search for information is often time-consuming, offering merely a black hole of empty space. To date, there has been no centralized source of this obscure information.

Researchers and reputable laboratories publish information which reaches a limited readership. This information may or may not reach the broad spectrum of practicing environmental professionals. Some laboratories create approaches of questionable import yet, due to extensive marketing, become recognized while others remain silent, unrecognized. Subsequently, out of frustration, some environmental professionals stylize their own methods, and others limit themselves to established, sometimes outdated techniques which rarely address the more complex issues which are becoming more prevalent in our technologically advancing society.

The term "environmental professional" is intended to be all inclusive of industrial hygienists, environmental scientists, and occupational health professionals. Information contained herein may be of interest and import to allergy specialists, scientific researchers, and public health officials as well.

It is the intent of this book to consolidate important advances which are established, seldom-used, and rarely acknowledged. The environmental professional facing complex nontraditional environmental issues can then make more informed decisions.

This book provides guidance to strategies, methodologies, interpretations, and limitations of heretofore decentralized, disjointed techniques. A broad spectrum of pertinent information is provided for the environmental professional to intelligently pick-and-choose methodologies which address unique scenarios.

Topics include a clarification of mold spore sampling and monitoring for unknown chemicals to a "how to" sample for *Legionella* and other airborne pathogens. There are approaches for determining the presence of a carcinogens as well as techniques for evaluating a broad spectrum of allergens in the air and dust samples. The tool box of methodologies includes forensic identification of dust as well as allergen characterization by immunoassay methods. Chemical analyses range from bag and canister sampling to several desorption methodologies. Methodologies are discussed in-depth and prominent experts in each field referenced. Information regarding commercial laboratories is accessible, and special sampling equipment is discussed in the appendices.

In summary, environmental professionals interested in advancing their personal knowledge and professional expertise will find this book an arsenal of information on the path to tomorrow's cutting edge. The contents of this book have been extensively researched. The information is easily referenced, and the text is clear and concise. This publication is a must have!

Kathleen Hess, MS, CIH

TABLE OF CONTENTS

Chapter 1
THE SEA OF UNKNOWNS

Man's environment is indeed a complex, ever-changing sea of chemical and biological confluences. Nature's medley of allergens and natural poisons is contributed to by man's ever-evolving creation of exotic substances. Yet, as more are added to the already overwhelming list of chemicals, identification of unknowns seems to have taken a back seat to progress.

Whereas the identity of an unknown must be suspect and its presence affirmed, many unknowns remain unknown; therefore, they do not exist. The frequency with which failed identification occurs within the environmental professions can only be speculative. It has been proclaimed, "The first person to cry wolf rules the consideration."

In 1982, an event involving a titanium oxide spill on the Bay Bridge in San Francisco, California, serves as a red herring story. A white powder was deposited by a transporter on the Bay Bridge, setting off a comedy of errors. The bridge was closed. A sample was extracted by an environmental firm, fully attired for the worst in "moon suits." Rumors culminated in the speculation that the material was an organophosphate. Several hours into the emergency, intensive analyses were being performed to identify an organophosphate. Still, the unknown remained "unknown." Only after rumor and speculation were discarded could the true culprit be identified, some twelve hours after the event. By this time, there were several hundred angry commuters who had long since parked their cars on the bridge. Rumors and hysteria have been the driving force for "looking in all the wrong places."

Allergies are typically associated with mold spores and pollen, and daily outdoor levels are typically provided by local news media. Yet, other substances may be implicated with allergies, and these are rarely addressed by the environmental professional. When allergies abound in an occupied office building where sampling has indicated low relative levels of mold spores and pollen, the complaints are often deferred as psychosomatic. Other allergens are frequently ignored or remain unknown.

Indoor air quality concerns have been nurtured by the news media, formaldehyde off-gassing is typically the ever-looming culprit. Subsequently, the environmental professional is obligated to address the formaldehyde levels even where symptoms and site history indicate otherwise. It is not uncommon for the environmental professional to be asked, in the same breath, to affirm that there are no "bad chemicals" present in the building when that which was sampled for was found not to be present. In one such instance, an environmental professional was told to sample for formaldehyde, carbon monoxide, carbon dioxide, and total organics in a building where the occupants were complaining of sewage odors. As the sampling strategy was driven by hearsay, not logic, the situation escalated into a political basketball with management attempting to block further sampling which may have addressed the "real issue."

Air pollution studies are typically based on that which is known or anticipated to be present in a given environment. Around industrial activities, the anticipated stack emissions are based on a composite of substances known to be exhausted, not on the complex composite of the reactions/mixing which may occur in the exhaust system or when exposed to a chemically complex atmosphere. These unknowns are rarely identified.

Many of the tools for identification and sampling of unknowns are discussed within this and succeeding chapters. The intent herein is to provide a screening process as well as several means whereby the environmental professional might expand one's technical tool box. This chapter deals with some basic screening approaches which are becoming widely accepted practices in limited niches while slowly becoming recognized in environmental hazardous waste management.

WASTE HAZARD CHARACTERIZATION AND BULK COMPONENT IDENTIFICATION[1]

Prior to disposal, a waste generator (e.g., manufacturer or property owner/recipient of midnight dumping) must characterize wastes either by performing expensive analytical procedures or applying knowledge of the waste due to known constituents and/or processes. Although the known components may undergo chemical alteration when mixed, some analyses may be eliminated (e.g., metals and pesticides would not be expected where solvents have been mixed).

On the other hand, without process information, identifying contents of containers with unknown constituents and no historic information may require full analytical consideration. To address unknown hazardous wastes, generalized background information is provided along with a discussion of a field test to characterize unknowns for initial management purposes.

Hazardous Waste Background

Under EPA Section 1004(5) Subtitle C of the Resource Conservation and Recovery Act (RCRA), a hazardous waste is defined as a solid waste (or combination of solid wastes) which because of its quantity, concentration, or physical/chemical/infectious characteristics may: (1) cause/significantly contribute to an increase in mortality or serious irreversible/incapacitating reversible illness; or (2) pose a substantial present/potential future hazard to human health or the environment (when the waste has been improperly treated, store, transported, disposed of, or mismanaged). Specific hazardous solid wastes are regulated, and their clarification/ identification is important.

"Solid waste" refers to any substance (solid, liquid, or contained gas) which is not excluded under 40 CFR 261.4 (e.g., some exclusions include domestic waste, sewage, and household wastes); material which has been abandoned (e.g., dispose of, burned, incinerated, accumulated, stored, or treated); and material which is considered inherently waste-like when recycled (e.g., waste from the production or use of pentachlorophenol).[1] If a material meets all the criteria for solid waste, it must also be one of the substance listed hazardous waste or exhibit at least one of the characteristics of a hazardous waste. The listed wastes are divided into three main categories:

- Hazardous wastes from nonspecific sources (F-code wastes; e.g., solvent wastes)
- Hazardous wastes from specific sources (K-code wastes; e.g., oven residue from production of chrome oxide green pigments)
- Commercial chemical product wastes — acutely hazardous (P- and U-codes; e.g., arsenic pentoxide)

Where a material is not listed, it may still meet the requirements for hazardous wastes. "Hazardous waste" characterization includes one of, or a combination of, the following hazards:

- Toxic
- Ignitable
- Corrosive
- Reactive

The "toxic" characterization is designed to identify wastes which are most likely to leach into the ground water and pose a threat to human health. The toxic substances which meet this criterion are limited to those identified in Table 1.1.

"Ignitability" is a physical hazard which poses a concern mostly during handling. To be classified as an ignitable, a solid waste must meet one or a combination of the following:

- A liquid that has a flash point of less than 140° F as determined by either a Pensky-Martens or Setaflash closed-cup test.

- A solid that is capable (under standard temperature and pressure) of causing fire through friction, adsorption of moisture, or spontaneous chemical changes and burns vigorously and persistently when ignited.
- An ignitable compressed gas as defined by the Department of Transportation (DOT).
- An oxidizer as defined by DOT.

A solid waste is "corrosive" if it meets at least one of two criteria. These are: (1) it is aqueous and has a pH of 2 or less, or of 12.5 or more; or (2) it is a liquid and corrodes steel at a rate of 6.35 mm or more per year as determined by the National Association of Corrosion Engineers. Solids are excluded.

Table 1.1 Toxicity Characteristic Leaching Procedure (TCLP) Constituents

Category/Chemical	Limit (mg/L)	Category/Chemical	Limit (mg/L)
PESTICIDES		SEMIVOLATILES	
Chlordane	0.03	o-Cresol	200
Endrin	0.02	m-Cresol	200
Heptachlor	0.008	p-Cresol	200
Lindane	0.4	1,4-Dichlorobenzene	7.5
Methoxychlor	10	2,4-Dichlorobenzene	0.13
Toxaphene	0.5	Hexachlorobenzene	0.13
		Hexachlorobutadiene	0.5
HERBICIDES		Hexachloroethane	3
2,4-D	10	Nitrobenzene	2
2,4,5-TP Silvex	1	Pentachlorophenol	100
		Pyridine	5
VOLATILES		2,4,5-Trichlorophenol	400
Benzene	0.5	2,4,6-Trichlorophenol	2
Carbon tetrachloride	0.5		
Chlorobenzene	100	METALS	
Chloroform	6	Arsenic	5
1,2-Dichloroethane	0.5	Barium	100
1,1-Dichloroethylene	0.7	Cadmium	1
Methyl ethyl ketone	200	Chromium	5
Tetrachloroethylene	0.7	Lead	5
Toluene		Mercury	0.2
Trichloroethylene	0.5	Selenium	1
Vinyl chloride	0.2	Silver	5

Excerpted from *Hazardous Waste: Identification and Classification Manual.*[1]

A "reactive" waste is one which has the capability to explode or undergo violent chemical change in a variety of situations. This characteristic identifies wastes which, because of their extreme instability and tendency to react violently or explode, pose a threat to human health and the environment at all stages of the waste handling process. Criteria are as follows:

- Instability and readiness to undergo violent changes.
- Violent reactions when mixed with water.
- Formation of potentially explosive mixtures when mixed with water.
- Generation of toxic fumes in quantities sufficient to present a danger to human health or the environment when mixed with water.
- Cyanide or sulfide bearing waste which generates toxic fumes when exposed to acidic conditions.
- Ease of detonation or explosive reaction when exposed to pressure or heat.
- Ease of detonation or explosive decomposition or reaction at standard temperature and pressure.
- Defined as a forbidden explosive, a Class A, or a Class B explosive by DOT.

Unknown Waste Fingerprinting[2,3]

From a spill on a highway to fields of drums containing unidentified wastes, identification of unknowns is like searching for a needle in a haystack. The line of logic is typically to: "Guess, then test." This approach is time consuming and expensive, oftentimes impractical.

Although several approaches have been attempted to provide a quick and easy process of elimination, commercially available identification kits have in recent years been refined. Initially, simple chemistry kits were developed to categorize waste. Although these kits were adequate in most instances, they were not foolproof. For instance, a typical kit would use pH paper to determine corrosivity, oxidizer paper to determine reactivity, and matches to determine ignitability. Yet, a high or low pH does not define all corrosives (e.g., phenol has a pH of 6.0 and is highly corrosive on skin), and oxidizer paper is negative with some of the stronger oxidizers (e.g., benzol peroxide and perchlorates). Then, one category remains unaddressed. Beyond the use of an assistant to sniff and/or taste the unknown, toxic substances are unidentified.

A refinement of the simplified test kits has led to a couple of field kits which are presently being used by fire fighters and hazmat responders to accomplish field tests during emergency and nonemergency waste encounters. They are increasingly being used by the environmental professional, and they have a particular appeal in complex, high quantity categorization/identification of unknown waste. Whereas a seasoned chemist may require less than 15 minutes, an untrained user (with kit and how-to manual in hand) may take 1 to 5 hours with most industrial waste. On the other hand, exotic substances, developed in small quantities within a laboratory, may require more time, more refined chemical and analytical protocols. The limitations of field chemistry are restricted to the less exotic, more widely used substances.

The commercially available HazTech® System's HazCat® Chemical Identification System is one of the more widely used kits. It has over 113 field tests and/or observations for identifying over 10,000 chemicals. The process is based on sound principles of basic chemistry and good old-fashioned common sense. The kit provides more than 60 chemicals along with necessary supplies, all of which are contained within a hand-carried plastic box which is easily managed by one person where testing can be performed on-site. See Table 1.2 for a generalized listing of some of the unknowns which are identified by the kit.

Besides liquid identification, the kit has been developed for solids and contained (or point source) gases. The decision trees are provided for both liquids and solids in Figures 1.1 and 1.2. Field support and training are available through the originators of the system.

Table 1.2 Substance Categories and Those Identifiable in the HazCat® Chemical Identification System

ORGANIC MATERIAL
- *Plants and Animals*
- *Medicines and Drugs*
  - Opium and cocaine
- Dyes (general identification)
- Chemicals
  - Organics
  - Inorganics
- Explosives
  - Hydrides, nitrates, nitrites, organic hydrazines, organic peroxides, and picric acid
  - Gun powder and periodates
  - Generalized less common exotic explosives
- *Military nerve and irritant gases*
- Pesticides
  - Organophosphates
  - Chlorinated pesticides
  - 2,4-D or 2,4,5-T
  - Carbamates
  - Thionated pesticides
  - Dioxin precursors
  - Metaldehyde
- Plastics
  - Thermoplastic
    - PTFE, CTFE, PVF, and FEP
    - ABS, acetal, cellulose acetate, cellulose propionate,

Table 1.2 (continued)

polystyrene, polyester, cellulose nitrate, and polyurethane
Nylon, polycarbonate, PPO, polysulfone, and PVC
Polypropylene
Polyethylene
Thermoset
DAP, melamine formaldehyde, phenol formaldehyde, and urea formaldehyde
Silicone and epoxy
Catalysts
Methyl ethyl ketone peroxide
Benzoyl peroxide
Soaps, detergents, and surfactants
Wood and wood products
Processed foods
Paints and shellacs
Denatured alcohol
Latex
Acrylic
Mineral spirits
Linseed oil in mineral spirits
1,1,1-Trichloroethane
Industrial oils, solvents, and production precursors
Polychlorinated biphenols

INORGANIC MATERIAL
Water
Silicates
Salts
Rocks/minerals
Acids
Oxidizing acids
Chromic acid
Organic acids
Hydrofluoric acid
Mineral acids
Acid gases
Alkali
Metallic and nonmetallic ionic compounds
Anionic compounds
Glass
Inorganic explosives
Asbestos

Excerpted from *Haz Cat Identification System Users Manual.* [4]

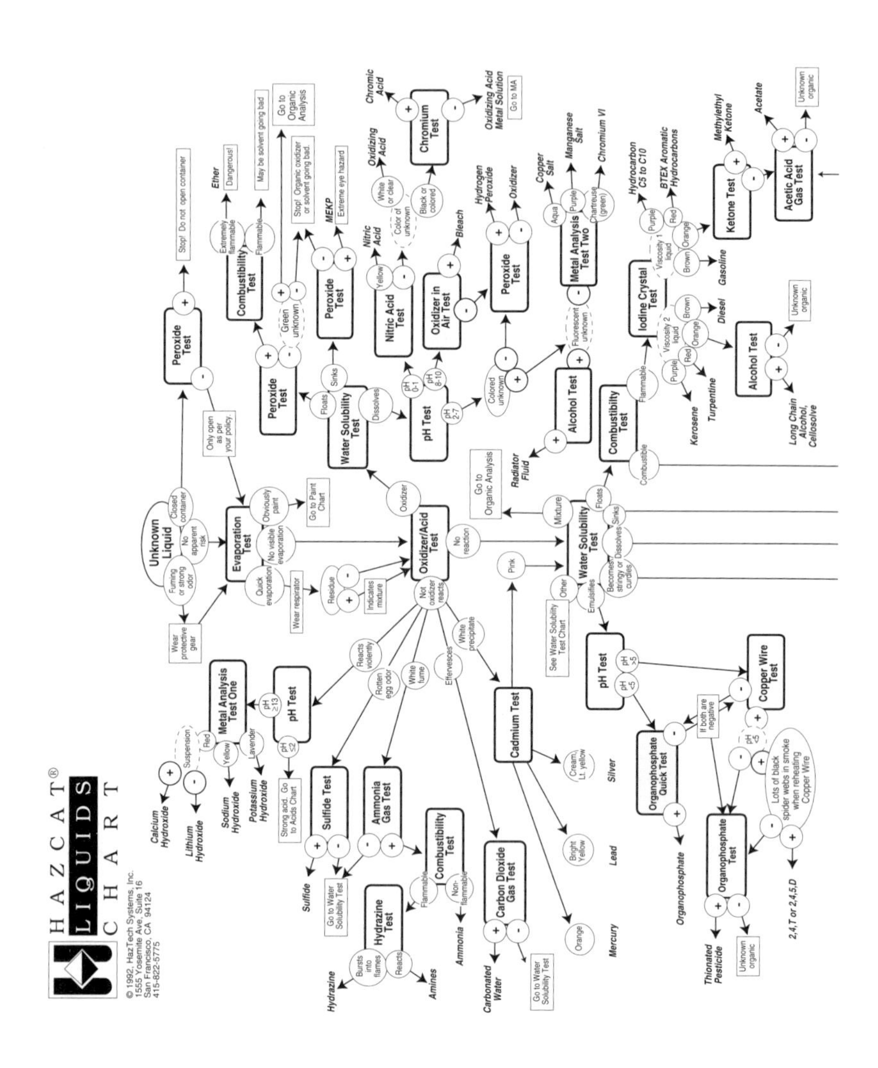

HAZCAT® LIQUIDS CHART
© 1992, HazTech Systems, Inc.
1555 Yosemite Ave, Suite 16
San Francisco, CA 94124
415-822-5775
Unknown Liquid
Fuming or strong odor
No apparent risk
Closed container
Wear protective gear
Only open as per your policy.
Peroxide Test
Stop! Do not open container
Evaporation Test
Obviously paint
Go to Paint Chart
No visible evaporation
Quick evaporation
Wear respirator
Residue
Indicates mixture
Peroxide Test
Combustibility Test
Extremely flammable
Ether
Dangerous!
Flammable
May be solvent going bad
Green unknown
Go to Organic Analysis
Stop! Organic oxidizer or solvent going bad.
Peroxide Test
MEKP
Extreme eye hazard
Water Solubility Test
Floats
Sinks
Dissolves
Oxidizer
Oxidizer/Acid Test
pH Test
pH 0-1
pH 8-10
pH 2-7
Nitric Acid Test
Yellow
Nitric Acid
Color of unknown
White or clear
Oxidizing Acid
Black or colored
Chromium Test
Chromic Acid
Oxidizing Acid Metal Solution
Go to MA
Oxidizer in Air Test
Bleach
Colored unknown
Peroxide Test
Hydrogen Peroxide
Oxidizer
Fluorescent unknown
Metal Analysis Test Two
Aqua
Copper Salt
Purple
Manganese Salt
Chartreuse (green)
Chromium VI
Alcohol Test
Radiator Fluid
Go to Organic Analysis
No reaction
Mixture
Water Solubility Test
Floats
Sinks
Dissolves
Becomes stringy or curdles
Emulsifies
Other
See Water Solubility Test Chart
Pink
Combustibility Test
Flammable
Combustible
Iodine Crystal Test
Viscosity 1 liquid
Viscosity 2 liquid
Purple
Hydrocarbon C5 to C10
Red
BTEX Aromatic Hydrocarbons
Orange
Brown
Gasoline
Diesel
Kerosene
Turpentine
Ketone Test
Methylethyl Ketone
Acetic Acid Gas Test
Acetate
Unknown organic
Alcohol Test
Long Chain Alcohol, Cellosolve
Unknown organic
Not oxidizer reacts
Reacts violently
pH Test
pH ≥13
pH ≤2
Strong acid. Go to Acids Chart
Metal Analysis Test One
Red
Yellow
Lavender
Suspension
Calcium Hydroxide
Lithium Hydroxide
Sodium Hydroxide
Potassium Hydroxide
Rotten egg odor
Sulfide Test
Sulfide
Go to Water Solubility Test
White fume
Ammonia Gas Test
Combustibility Test
Flammable
Non-flammable
Ammonia
Hydrazine Test
Bursts into flames
Hydrazine
Reacts
Amines
Effervesces
Carbon Dioxide Gas Test
Carbonated Water
Go to Water Solubility Test
White precipitate
Cadmium Test
Cream, Lt. yellow
Silver
Bright Yellow
Lead
Orange
Mercury
pH Test
pH >5
pH <5
Copper Wire Test
If both are negative
Organophosphate Quick Test
Organophosphate
pH <5
Lots of black spider webs in smoke when reheating Copper Wire
Organophosphate Test
2,4,T or 2,4,5,D
Thionated Pesticide
Unknown organic

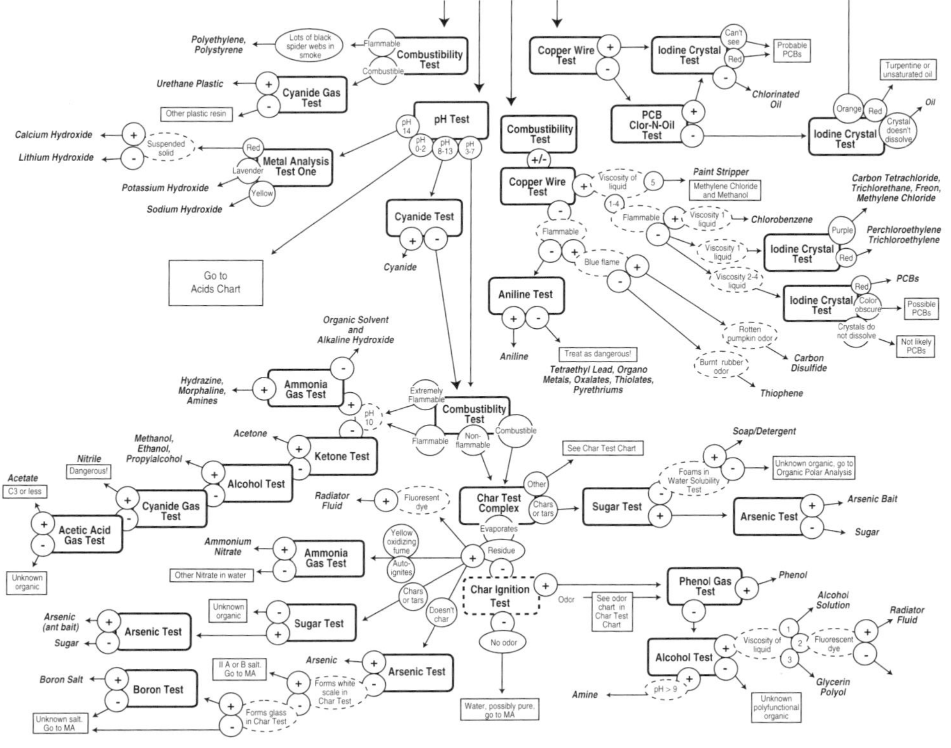

Figure 1.1 "Decision tree" for unknown liquids. (Courtesy of HazTech® Systems, Inc., San Francisco, CA)

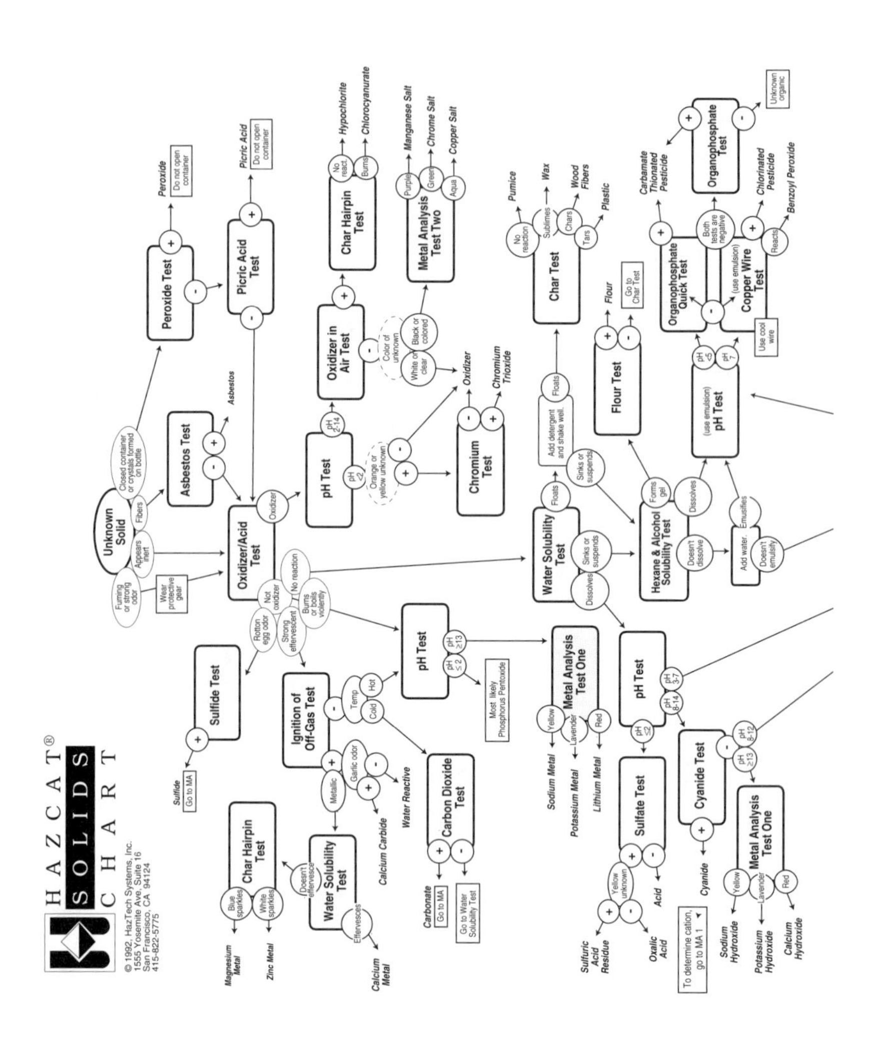
HAZCAT® SOLIDS CHART
© 1992, HazTech Systems, Inc.
1555 Yosemite Ave, Suite 16
San Francisco, CA 94124
415-822-5775
Unknown Solid
Fuming or strong odor
Wear protective gear
Appears inert
Fibers
Closed container or crystals formed on bottle
Asbestos Test
Asbestos
Peroxide Test
Peroxide
Do not open container
Picric Acid Test
Picric Acid
Do not open container
Oxidizer/Acid Test
Oxidizer
pH Test
pH 2-14
pH <2
Orange or yellow unknown
Chromium Test
Chromium Trioxide
Oxidizer
Oxidizer in Air Test
Color of unknown
White or clear
Black or colored
Char Hairpin Test
No react.
Burns
Hypochlorite
Chlorocyanurate
Metal Analysis Test Two
Purple
Green
Aqua
Manganese Salt
Chrome Salt
Copper Salt
Not oxidizer
No reaction
Rotton egg odor
Strong effervescent
Burns or boils violently
Sulfide Test
Sulfide
Go to MA
Ignition of Off-Gas Test
Temp
Hot
Cold
Metallic
Garlic odor
Calcium Carbide
Water Reactive
pH Test
pH ≤2
pH ≥13
Most likely Phosphorus Pentoxide
Carbon Dioxide Test
Carbonate
Go to MA
Go to Water Solubility Test
Water Solubility Test
Doesn't effervesce
Effervesces
Calcium Metal
Char Hairpin Test
Blue sparkles
White sparkles
Magnesium Metal
Zinc Metal
Metal Analysis Test One
Yellow
Lavender
Red
Sodium Metal
Potassium Metal
Lithium Metal
Water Solubility Test
Floats
Sinks or suspends
Dissolves
Add detergent and shake well.
Floats
Sinks or suspends
Char Test
No reaction
Sublimes
Chars
Tars
Pumice
Wax
Wood Fibers
Plastic
Hexane & Alcohol Solubility Test
Forms gel
Dissolves
Doesn't dissolve
Flour Test
Flour
Go to Char Test
Add water.
Emulsifies
Doesn't emulsify
(use emulsion) pH Test
pH <5
pH 7
Organophosphate Quick Test
(use emulsion) Copper Wire Test
Use cool wire
Both tests are negative
Reacts
Chlorinated Pesticide
Benzoyl Peroxide
Organophosphate Test
Carbamate Thionated Pesticide
Unknown organic
pH Test
pH ≤2
pH 8-14
pH 3-7
Sulfate Test
Yellow unknown
Sulfuric Acid Residue
Oxalic Acid
Acid
Cyanide Test
Cyanide
pH 8-12
pH ≥13
Metal Analysis Test One
Yellow
Lavender
Red
Sodium Hydroxide
Potassium Hydroxide
Calcium Hydroxide
To determine cation, go to MA 1

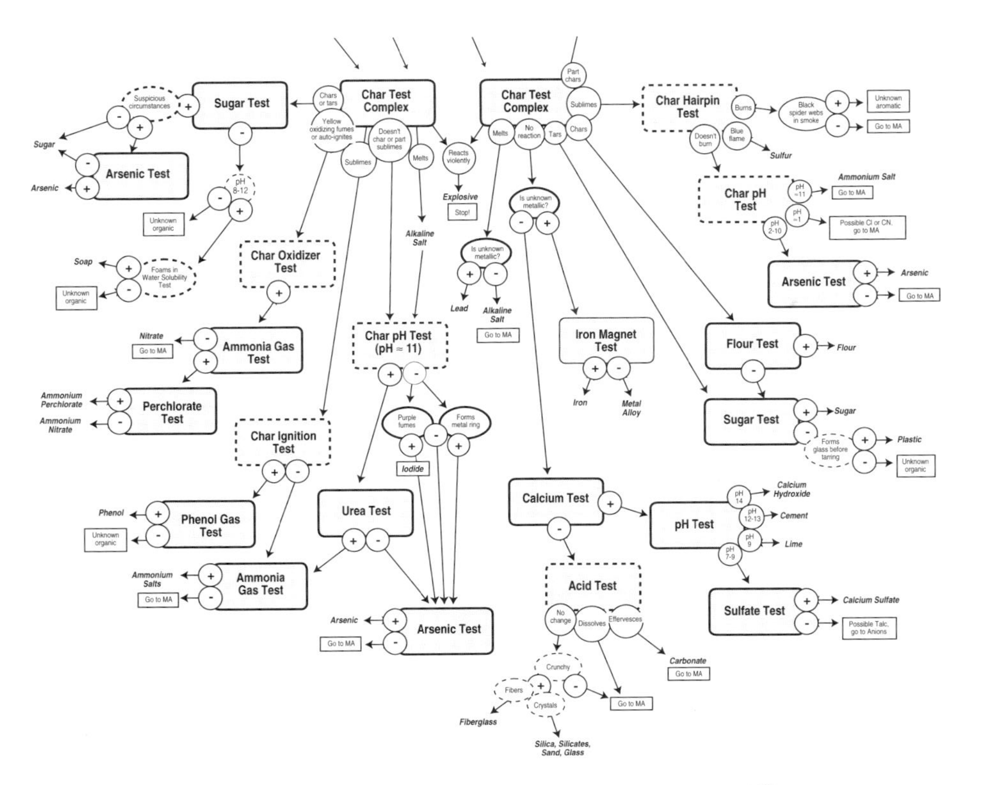

Figure 1.2 "Decision tree" for unknown solids. (Courtesy of HazTech® Systems, Inc., San Francisco, CA)

Opponents express a concern that the kit could not possibly address more than singular unknowns. Yet, proponents state that not only can most substances be identified in complex mixtures, but components may be detected at levels as low as parts of contaminant per million parts of sample (ppm). Most of the inorganics are identified, and the organics are categorized down to functional groupings at a minimum (e.g., chlorinated hydrocarbons).

With limited knowledge, the user may easily categorize a substance for purposes of handling and shipping. With a little more experience and skill, one may identify individual components by functional group (e.g., chlorinated aliphatic), if not by name, and the substances listed for TCLP analyses may be screened.

Whereas the EPA analytical methods for addressing unknowns involve the "process of elimination" regarding known/listed substances, the Chemical Identification System attempts to identify regardless of whether a substance is a known hazardous material or not. Complete identification is particularly helpful when addressing emergency spills where the news media may have become involved. A news statement indicating an "unidentified nonhazardous" fails to inspire confidence.

Unknown Gases[5,6]

Due to the low levels of unique airborne components, unknown gases are not as easily addressed as the liquids and solids. Presently, there are several commercially available kits which tout the ability to identify unknowns in the air. Yet, even these kits are limited. Where these kits depart, the methodologies presented in the succeeding chapters should fill in the gaps.

Most kits involve a series of detector tubes with an expanded ability to screen for a broad spectrum of substances, with "limited levels of detection." See the decision tree in Figure 1.3. Due to the limits of detection, the detection process has limited use in indoor air quality evaluations where the levels of contributing unknowns are not likely to exceed the stated detection limits. Thus, an unknown will remain undetected. For instance, the first detector tube in the decision process is the Polytest which is based on easily oxidized compounds. Some of the components identified are provided in Table 1.3. In indoor air quality surveys, many organics are found in levels below 30 ppm which is an extreme level for indoor air. Typically, a year after occupancy, the total organics are at or below 1 ppm.

The most feasible use for the gas detector kits is where the unknown is contained (or the evolutionary point source is known) and concentrated levels exist. A compressed gas cylinder may require content identification which is also part of the HazMat® Chemical Identification Kit. A gas may be evolving from a point source in the ground, or a gas may have been generated and contained upon mixing of two unknowns.

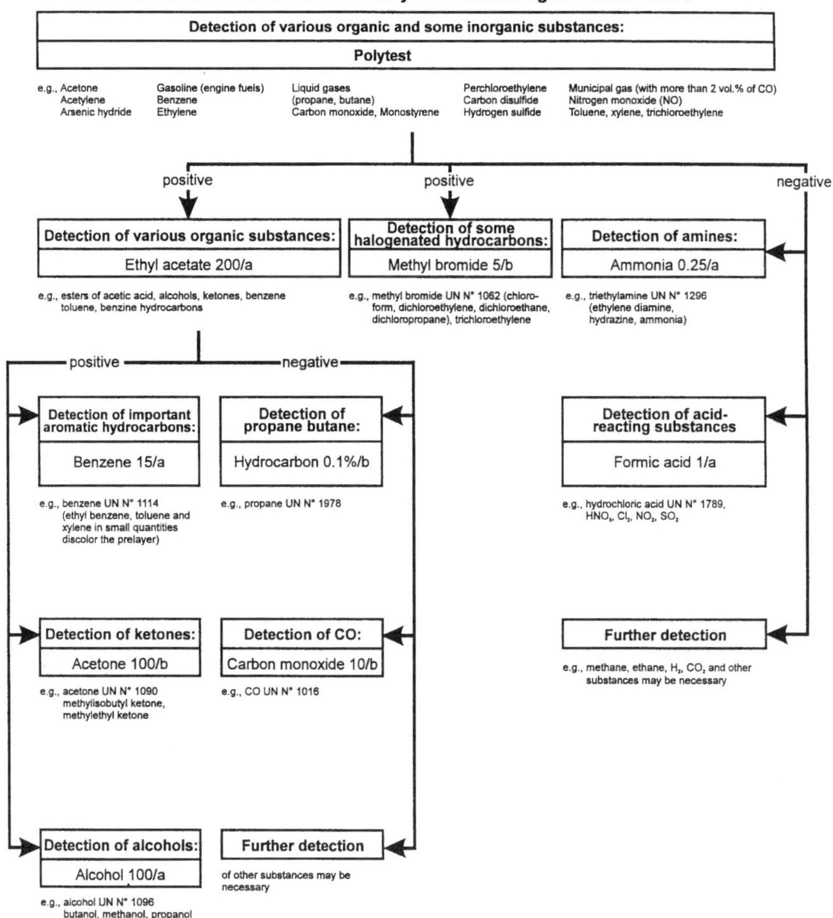

Figure 1.3 Draeger "decision tree" for unknown gases. (Courtesy of National Draeger , Pittsburgh, PA)

Table 1.3 Substances Detected by Polytest Tube

Substance	Lower Limit of Detection
Acetone	2000 ppm
Acetylene	10 ppm
Arsine 1 ppm	
Benzene	50 ppm
Butane	100 ppm
Carbon disulfide	1 ppm
Carbon monoxide	5 ppm
Ethylene	50 ppm
Hydrogen sulfide	2 ppm
Nitrogen oxide	50 ppm
Octane	10 ppm
Perchloroethylene	20 ppm
Propane	100 ppm
Styrene	10 ppm
Toluene, xylene	10 ppm
Trichloroethylene	50 ppm

Not detected: Methane, ethane, hydrogen, and carbon dioxide.
Excerpted from the *Draeger Detector Tube Handbook.*[5]

The Draeger Polytest detector tube produces a color change when any of these substances are present at the given detection limit only, but it does not indicate amount. A negative on the Polytest goes to detection of amines and detection of acid-reacting substances. When all else has failed, there are over 150 different detection tubes available for purchase. Thus, the feasibility and efficacy of this process for detection of low levels of an unknown should be challenged.

Unless a substance is suspect as being present in levels which exceed the acceptable occupational limits, a negative finding can only be interpreted as that substance (or group of substances) not being present at elevated levels. The colorimetric detector tubes do not negate the existence of low level toxins. They only negate the existence of toxic substances at levels in excess of acceptable work exposure limits.

These methods are best used in an emergency where an unknown toxic gas is contained or evolving from a known source at high levels. Extensive air sampling and expensive laboratory analyses may be deferred in emergencies where time often becomes the overriding factor.

Another method is available which makes use of the colorimetric detector tubes while providing some additional information, once again, on high levels. This other method is the HazCat® Atmospheric Test Sequence. This approach requires direct reading instrumentation, odor detection, pH test paper, and a logic "decision tree" in unison with the colorimetric detector tubes. Thus, flammable gases are added to the list of toxic substances, along with carbon monoxide, oxidizers, pesticides, and a broad spectrum of acids.[6] See Figure 1.4.

ALLERGENS

Typically, less than 20 percent of any given adult population will have predictable, on-going allergies. When the percentage of allergies in a population become elevated, excessive levels of one or more allergens becomes evident. The higher the percentage of allergy sufferers, the greater the probability of increased allergens. Yet, it is not always clear as to what allergen is the culprit. The culprit is an unknown!

When dealing with unknown allergens, all options should be considered. Typically, the mold spores are the primary suspect while other allergens are overlooked. Thus, the possibilities are presented herein so the environmental professional will have a broader perspective of potential causative agents. Environmental allergens for which procedures for identification and evaluation exist include the following categories:

- Plant pollen
- Mold spores
- Bacterial spores
- Animal allergens
- Some organic chemicals (rarely implicated in large populations)

The concern for allergens is associated with outdoor environments, indoor air quality office spaces, and indoor residential environments.

INDOOR AIR QUALITY

Internationally, indoor air quality complaints have become a frequently discussed office building epidemic. In a majority of the cases, the cause is assigned to poor ventilation without the specific agents being identified. It should be noted, also, that some environmental professionals feel that the problems are psychosomatic, once again, due to the inability to isolate any one of or group of substances which could be implicated. Many of the complaints involve symptoms similar to those of allergies. Some are associated with illness. Yet, most problems are wholly associated with a specific building or area(s) of a building.

Due to the complex nature of unknowns, even the most exhaustive of searches cannot possibly include all possibilities. For this reason, routine basics are addressing a mere "tip of the iceberg." A more expansive investigation

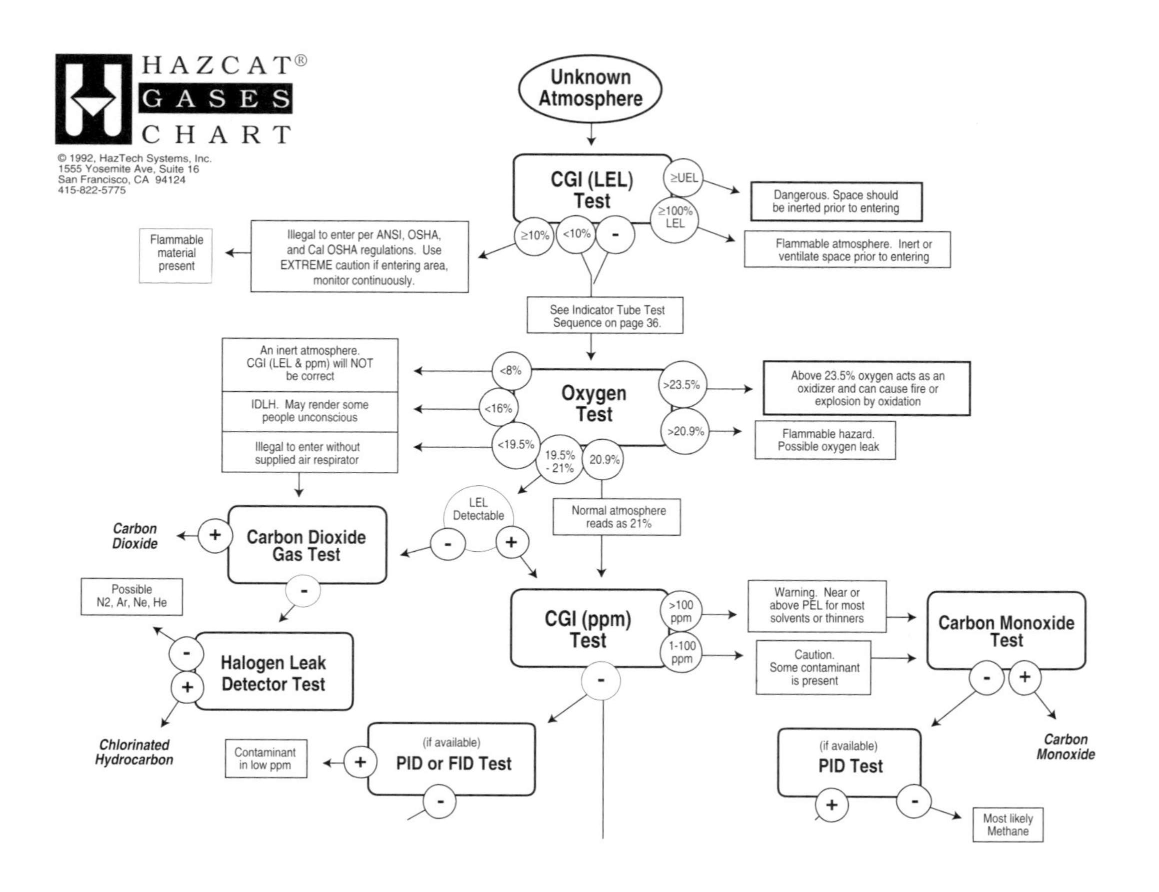
HAZCAT®
GASES
CHART
© 1992, HazTech Systems, Inc.
1555 Yosemite Ave, Suite 16
San Francisco, CA 94124
415-822-5775
Unknown Atmosphere
CGI (LEL) Test
≥10%
<10%
-
≥UEL
≥100% LEL
Dangerous. Space should be inerted prior to entering
Flammable atmosphere. Inert or ventilate space prior to entering
Illegal to enter per ANSI, OSHA, and Cal OSHA regulations. Use EXTREME caution if entering area, monitor continuously.
Flammable material present
See Indicator Tube Test Sequence on page 36.
Oxygen Test
<8%
<16%
<19.5%
19.5% - 21%
20.9%
>23.5%
>20.9%
Above 23.5% oxygen acts as an oxidizer and can cause fire or explosion by oxidation
Flammable hazard. Possible oxygen leak
An inert atmosphere. CGI (LEL & ppm) will NOT be correct
IDLH. May render some people unconscious
Illegal to enter without supplied air respirator
Normal atmosphere reads as 21%
LEL Detectable
-
+
Carbon Dioxide Gas Test
+
-
Carbon Dioxide
Possible N2, Ar, Ne, He
Halogen Leak Detector Test
-
+
Chlorinated Hydrocarbon
CGI (ppm) Test
>100 ppm
1-100 ppm
-
Warning. Near or above PEL for most solvents or thinners
Caution. Some contaminant is present
Carbon Monoxide Test
-
+
Carbon Monoxide
(if available) PID or FID Test
+
-
Contaminant in low ppm
(if available) PID Test
+
-
Most likely Methane

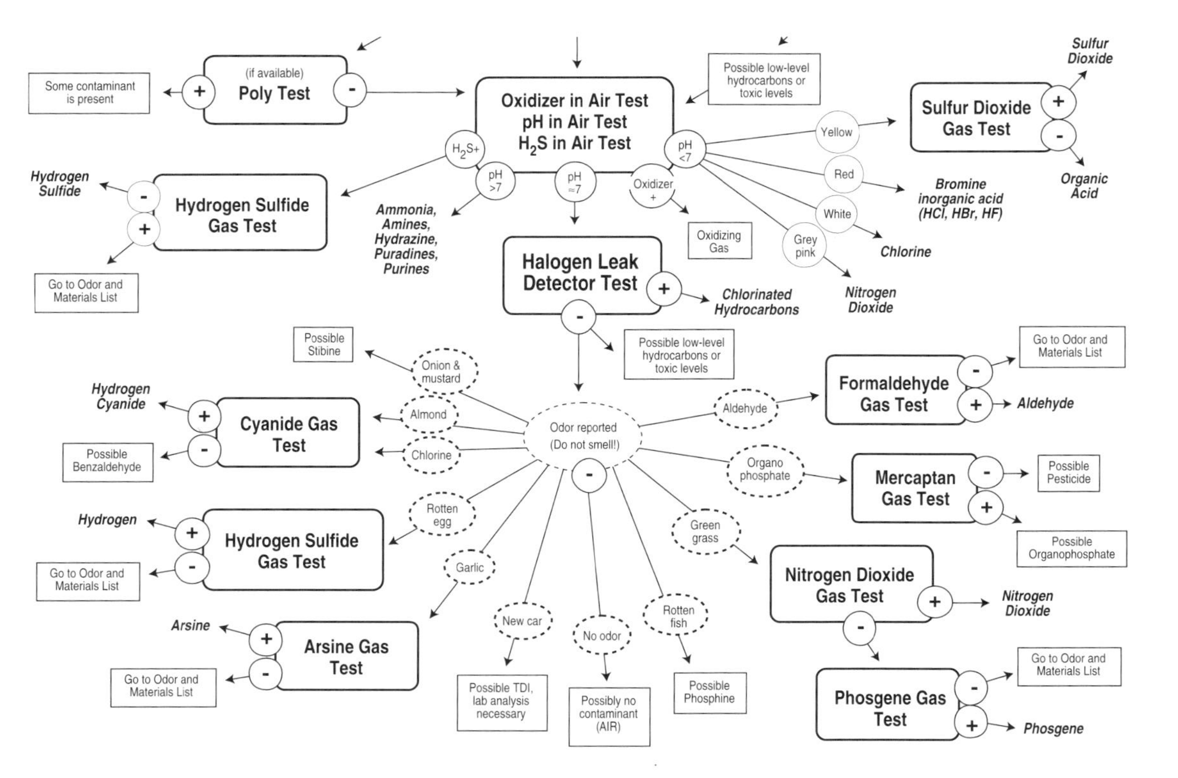

Figure 1.4 HazTech® "decision tree" for unknown gases. (Courtesy of HazTech® Systems, Inc., San Francisco, CA)

has in some instances disclosed relevant information with corrective actions which have not required expensive make-up/fresh air conditioning and addition a building ventilation system. Routine sampling includes sampling for carbon monoxide, carbon dioxide, total hydrocarbons, and formaldehyde. Additional categories posed herein as situation dependent are:

- Allergens
- Mycotoxins
- Endotoxins
- Total volatile/semivolatile organic compound identification
- Pathogenic microbes
- Microscopic particulates
- Carcinogens/mutagens
- Product emissions

AIR POLLUTION

The complexities of air pollution are complicated by the multiple exhaust stack emissions, car emissions, nature's supermarket of airborne substances, and photochemical reactions. Known industrial processes create by-products which may or may not be known. These are exhausted, along with other by-products which may or may not be known. Exhausted into the atmosphere, they may or may not react with other industrial emissions as well as with auto emissions. Nature's own bounty of metabolic and reproductive processes contribute to the environmental ambrosia. Then, the brew of unknowns undergoes ongoing photochemical alterations. This ever-changing complex of unknowns becomes our life force.

The Environmental Protection Agency requires routine stack sampling for known emissions, and contributions are locally controlled. The known industrial emissions are monitored at the stack. Contributing known auto emissions are figured into the pollution scheme. The final witches' brew of pollutants remains unknown. A means for identification and quantitation of some of these unknowns is provided herein under the following categories:

- Allergenic mold spores and pollen
- Total volatile/semivolatile organic compound identification
- Microscopic particulates
- Carcinogens/mutagens

REFERENCES

1. Wagner, Travis P. *Hazardous Waste: Identification and Classification Manual.* Van Nostrand Reinhold, New York, 1990.
2. Turkington, Robert. HazCat® usage and special problems. [Oral communication] HazTech® Systems, Inc., San Francisco, California, December 1995.
3. Turkington, Robert. *HazCat ®Chemical Identification System User's Manual.* HazTech® Systems, Inc., San Francisco, California, 1994.
4. Ibid. p. 20
5. Draeger. *Draeger Detector Tube Handbook.* National Draeger, Inc., Pittsburgh, 8th Edition, 1992.
6. Turkington, Robert. *HazCat® Atmospheric Test Sequence.* HazTech® Systems, Inc., San Francisco, California, 1996.

Chapter 2

ALLERGENICITY, PLANT POLLEN, AND MOLD SPORES

Our universe is a sea of foreign substances capable of eliciting an allergic response in our immune systems, fortifying the body against potentially damaging invaders. This defense response of our immune systems may result in clinical symptoms of allergenicity which can be debilitating, the central focus in the lives of many allergy sufferers.

In terms of lives impacted, allergenic substances are at the top of the list. There are twenty times more people affected by allergens than cigarettes and one thousand more than chemical pollutants. Yet, their importance is often diluted. This is due to lack of symptom severity (e.g., rare occurrence of death) and to the acceptance that environmental controls may be a study in futility.

Allergens are everywhere, except in the most restrictive of environments (e.g., an environmentally controlled, filtered bubble enclosure). The response of one's body to their presence occurs more aggressively when specific allergens are elevated. Even with reduced levels, some people are more sensitive than others. For the more sensitive individual, the mere presence of an allergen at any dose level could pose a hostile assault on the body's defense mechanisms, possibly resulting in death. In some cases, as with the antibiotic mold *Penicillium*, one man's safety net may be another man's coffin. The concern may be life-threatening.

As environmental professionals, the burden of understanding and interpreting the cause and effect of various doses is becoming ever more important. Some allergists have dared to allege a given pollen to be the "probable source" of a patient's allergies, basing the claim on seasonal norms. Later, that allergist finds that pollination of the suspect plant had yet to occur. Pollination had not followed the norm. Thus, the allergist's speculation was faulty due to insufficient information.

Local newspapers typically report pollen and spore counts. The counts are provided by any of a number of sources, oftentimes by a local allergy clinic, and each reporting entity uses a different method for sampling, identifying, and interpreting the results. Although these various methods are designed with good intent, until recently, there had been no movements to standardize the methodologies.

Physicians/allergists and allergy sufferers use newspaper reports to assist in diagnosing allergy problems. Yet, of all the thousands of local newspapers providing this service, there are less than a hundred sites performing according to a standardized procedure.

Strategy and environmental relationships are frequently overlooked as well. That which represents the outdoor pollen and spore allergens does not necessarily hold true for the indoor environments. There have been reported findings within indoor office and household environments where the pollen count is not only high, but the type of pollen identified is no longer found in the outdoor air. A possible reason for this occurrence is entry and subsequent retention within a poorly filtered air handling system to a building during the "allergy season." The pollen builds up and is recycled with the supplied air.

Relevant to this chapter, the mechanisms of airborne plant allergens are discussed. Symptoms are elaborated upon. Sources are described, and a spectrum of sampling methods are explained.

ALLERGENIC MECHANISM

The word allergy, from the Greek allos ("other") and ergeon ("action"), was coined in 1906 by Baron Clemens von Pirquet, who recognized that the introduction of a foreign substance into a tissue can alter the tissue's capacity to react to subsequent encounters to the same substance. This altered response, more appropriately referred to as hypersensitivity, can be either protective or, in extreme cases, harmful (e.g., anaphylactic shock). Allergies are an exaggerated response to low dose challenges from a substance "recognized as foreign."

The mechanism involves an initial challenge to a foreign substance (usually large doses), referred to as an antigen. The body responds to the challenge to form a specialized response team of antibodies to fend off subsequent challenges after the individual has been sensitized. The intent is to promote the elimination or destruction of undesirable material which has been identified by the individual's own unique immune system.

Antigens generally have a high molecular weight, and the higher the molecular weight, the more likely the foreign material will be recognized as undesirable. The low end for proteins is 10,000 daltons, and the low end for polysaccharides is in excess of 50,000 daltons.

The more different the antigen, the stronger the response. Plant proteins are stronger antigens for humans than are monkey proteins. Yet, this rule sometimes fails. On occasion, the immune system gets out of control. An individual's own body proteins may become unrecognizable by its own immune system and be attacked by that same mechanism which is meant to protect the body.

An antigen does not have to be introduced into the circulatory system or lungs in order to be tagged for destruction and elimination. It may also be introduced through other routes including, but not limited to, undigested foods, male sperm, and mother's milk.

Allergies develop when the immune system recognizes a foreign substance, responds to eliminate or destroy it, and continues to be challenged by the same substance. Repeated or continued exposures result in continued allergy symptoms. A reduction of levels helps. But exposures must then be either eliminated from the environment; the individual must relocate to avoid further exposures; or the individual must undergo costly medical treatment in order to remedy the discomfort.

Table 2.1 Airborne Causes of Allergic Reactions

General	Specific Antigen Types
pollen grains	weeds, grasses, and trees
spores	fungal and bacterial
body parts of animals	dander (e.g., cat or dog epithelial cells) feathers insect parts (e.g., dust mites)
industrial products	isocyanates (e.g., polyurethane) formaldehyde (e.g., urea formaldehyde resins) insecticides containing pyrethrums

In extreme cases, an individual's own immune system attacks that same individual which it was meant to protect. Auto immune diseases include rheumatic fever, rheumatoid arthritis, and systemic lupus erythematosus. On the other hand, auto immune deficiency syndrome (AIDS) is the opposite. One's immune system fails to protect and defend from foreign substances, allowing foreign substances to attack the unprotected cells and tissues of the patient. Immune suppressed patients are particularly susceptible to infection by some of the fungi which are merely allergenic to those with normal immune systems.

The foreign substance must possess molecular complexity. For instance, synthetic polymers are nonantigenic. Polystyrene, Teflon, nylon, and Saran are all not normally considered antigens despite their high molecular weight. Foreign substances must have surface molecular complexity to be recognized as an allergen.

Some materials which are simple in molecular structure, however, may be attached to more complex molecules to alter the identity of the previously identifiable molecule. These simpler molecules are referred to as haptens. Haptens connect with more complex molecules, allowing the formerly nonallergenic substance to illicit an allergic reaction. Some examples include toluene diisocyanate and vinyl chloride.

ALLERGY-RELATED SYMPTOMS

In some instances, the environmental professional must diagnose a problem based on symptoms and environmental conditions. Oftentimes, descriptions are over-inflated, confused, and fear driven. Textbook symptoms become difficult to follow if not understood in depth.

The complaintant describes a scratchy throat and itchy, light-sensitive eyes. A scratchy throat may be due to irritation caused by sinus drainage. Itchy eyes may result from hay fever irritation, and bright light contributes to eye sensitivity associated with allergic conjunctivitis. Yet, some of these symptoms are not always allergy-related. The predominant symptoms, observed in environmental situations, are described within this section in order of prevalence within the population.

Allergic Rhinitis

Allergic rhinitis (sometimes referred to as allergic rhinoconjunctivitis) is the most commonly encountered of all allergy conditions. Allergic reactions are characterized by nasal congestion and sneezing which are, in turn, associated with irritation of the throat, eyes, and ears.

Nasal Congestion

Nasal congestion may result in breathing difficulties due to blockage of the nasal passages. Congestion may also result in a watery, blood-flecked discharge either from the nose and/or back of the throat. Throat irritation may be described as a "tickle" by some allergy sufferers. Drainage and throat irritation provoke coughing and other indirectly related problems.

When nasal blockage occurs, allergy sufferers tend to breath through their mouth and frequently to lose sleep. The problem thus culminates in fatigue and irritability. Although these symptoms are associated with allergies, it is important that the environmental professional be aware that they may be the result of other problems which may require a physician.

Not all nasal discharges are caused by allergies. Simple colds or virus infections may simulate allergic disease. Allergic disease is not normally accompanied by the presence of fever, general aching, and a yellow or green nasal discharge. Where exposed to unusually high levels of an allergen, an allergy sufferer may develop a systemic reaction where general aching is the primary symptom. Yellow or green nasal/throat discharges are generally associated with viral or bacterial infections, and an infection lasts about two weeks. An allergic condition comes and goes with environmental changes.

Another cause of congestion which should not be confused with allergic rhinitis is that which is caused by a tumor obstruction. Nasal blockage may result from polyps—nonmalignant growths of sinus and nasal tissue filled with fluid, which may or may not be the result of allergy-caused post nasal drip. If the blockage persists unabated for several days, both loss of smell and nasal/sinus infections may result.

Eye Pain

Allergic conjunctivitis can induce a painful, burning condition known as "pink eye." Bright light often adds to the discomfort, leading the individual to appear to be afraid of light. Severe watering and squinting may occur.

Inner Ear Pain and Hearing Loss

Allergic rhinitis is associated with extensive fluid leakage into the middle ear. This fluid buildup causes an aching pain that can temporarily diminish hearing. If untreated, the condition can cause fever and increased pain, and lead ultimately to rupture of the eardrum.

Hay Fever

When symptoms are seasonal, allergic rhinitis is referred to as "hay fever." Such allergens include pollen and mold spores. Itching of the nasal passages is a common symptom associated with pollen and mold spores. In an attempt to relieve the discomfort, a sufferer frequently performs nose-rubbing rituals and facial contortions. Itching of the roof of the mouth and inner corner of the eyes may further contribute to demonstrations of distress.

Sinus Headache

Sinus obstruction results in area related headaches. The sinuses have outlets or air spaces through the nasal passages. These air spaces are located near the nose, under the cheeks, and above the eyes. When the spaces become obstructed, buildup of air or fluid leads to increased pressure, resulting in pain. This pain may range from a dull discomfort to a sharp, steady ache in the areas involved (e.g., above the eyes). This is referred to as a sinus headache, or "sinusitis."

Allergic Asthma

Asthma generally results when cold air, pungent odors, viral infections, aspirin and related drugs, inert dusts, or allergens cause hyperactivity of the airways with narrowing of the air passages. This is translated into a "tightening sensation of the chest and breathing difficulties associated with coughing and wheezing. Movement of the chest-wall muscles may be felt as a heavy weight, a constriction of the lungs. If complications involve the pneumothorax, rib fracture, or pneumonia, chest tightness becomes pronounced and painful. Breathing could become excruciating and resemble symptoms of pleurisy.

Strenuous exercise may also heighten the severity as may increased exposure levels. In severe episodes, the sufferer may also experience wheezing and mental dullness due to reduced oxygen inspired/delivered to the brain.

If a person has been relatively free of asthma until exposure to animals, cut grass, or a virus, and if the exposure is brief, the attack should subside with proper treatment. It may even subside spontaneously.

Allergic Dermatitis

Although acute symptoms of allergic dermatitis are not singularly diagnostic, they characteristically involve itching, redness, swelling, and a scaly rash.

Itching can be intense and may lead to what is known as "weeping" lesions, caused by the serum oozing from the underlying small blood vessels. Typical areas of the body for this to happen are the cheeks, the creases behind the ears, and at the bends of the arms and legs.

Commonly called eczema or atopic dermatitis, the rash can spread enough to become disabling. During the healing stage, the affected skin thickens, becomes dry and cracks. Some bleeding may also occur during this stage.

A local infection that takes the form of skin boils is serious and should be considered a threat to the comfort of the individual. As their immune system has yet to develop completely, infants and young children may not demonstrate as severe a reaction as an adult. They may simply develop a red, flat, scaly rash without the annoyance experienced by older individuals.

Other Allergy-related Illnesses

Some rare forms of allergy-related illness occur in less than one percent of the United States population. Awareness concerning some of their symptoms may be relevant to an overall investigation of environmental impact of airborne allergens. These are discussed herein.

Allergic Bronchopulmonary Aspergillosis

Allergic bronchopulmonary aspergillosis is, as the name implies, an allergic disease which involves exposures to the fungus *Aspergillus. Aspergillus fumigatus* is generally the species implicated. This is a disease involving invasion of the mucous lining of the air passages within the lungs. Affected individuals may experience a persistent cough associated with sputum, asthma-like symptoms, fever, and chest pain. Viable spores are the initiator, and, once invasion has occurred, the disease is progressive until treatment is administered.

Hypersensitivity Pneumonitis

Hypersensitivity pneumonitis, or "allergic alveolitis," generally involves an occupationally-related inflammation of the alveoli and bronchioles of the lungs. The acute disease, pursuant to extremely high exposures, is characterized by breathlessness, chills, and fever. Fever may be as high as 104° F, and, within four to six hours, symptoms progress into muscle aches and moodiness. Chest X-rays may show diffuse nodular shadows which are predominately found in the lower portions of the lungs.

Subacute symptoms occur upon repeated exposures and appear as chronic bronchitis. This phase is characterized by repeated exposures and is characterized by recurring cough, breathlessness, weight loss, and malaise. Chronic exposures to low levels of antigenic material may be demonstrated by less obvious symptoms. The allergy sufferer may only experience breathlessness and weight loss, but the chest X-rays will reveal progressive scarring. Pulmonary function tests will show declining functional lung volume. Diagnostic confirmation of the various stages of the disease generally requires the aid of a physician.

Common occurrence of the disease is typically to bird handlers and to farmers in the southeast. Disease is generally associated with a known source and is region dependent.

Hives

Hives, or "urticaria," is generally associated with food or drugs. On rare occasions is may be associated with animal allergens, skin contact with the mold *Penicillium*, and *Hevea brasiliensis* latex.

Raised bumps may appear anywhere on the body. These may range in size from that of a small pea to large sections of the body and are often accompanied by red haloes and itching. Each lesion may last only a few hours, but new ones can appear at frequent intervals, and itching may become so intense that it is difficult to perform simple tasks or sleep.

Contact Dermatitis

Contact dermatitis involves an itchy, red rash which is generally confined to the site of exposure. Scratching may lead to blistering. Although the areas impacted are commonly exposed skin surfaces, exposures may result from transference of allergen from one place to another (e.g., scratching unexposed skin surfaces with contaminated fingernails). The most common exposures are to oils from poison ivy and poison sumac which if burned may become airborne or are spread by contact with weeping sores. Removal from and avoidance of the allergen will, in most cases, provide relief within two weeks if the affected areas are left alone.

PLANT POLLEN AND MICROBIAL SPORES

There are over 100,000 different species of fungi and in excess of 350,000 species of plants. Many of these are potentially allergenic due, in large part, to the nature of the outer protective surface of the plant pollen and microbial spores.

In pollen allergenicity, plants have historically been characterized into three groups. These are trees, grasses, and weeds. Trees and grasses typically have distinct geographic distributions. The ornamental trees and desirable grasses are concentrated around population areas. Sometimes landscaping involves the placement of allergen-producing shrubs around population centers. Weeds grow in the wild. Although many of the common allergenic weeds were introduced to the United States as food plants or medicinal herbs, they are "generally" considered undesirable and are not cultivated.

Fungi include single celled yeast, filamentous molds, and multicellular mushrooms. The latter two are responsible for the creation of a majority of the airborne allergenic spores. Possessing a hard chitin or polysaccharide exterior covering, mold spores are typically resistant to drying, heat, freezing, and some chemical agents. Some bacteria also produce a protective outer covering under stress conditions (e.g., thermophilic actinomycetes).

Allergenic spores and pollen have been known to be transported by high winds as far as 1,500 miles, and it is possible to find them a hundred miles from their point of origin.[1] If one were to perform a contour map showing levels at various points from a source, they will find the highest concentrations close to the source, diminishing with distance and impacted by wind direction and velocity and volume of pollen produced at the source. When diagnosis and routine monitoring are performed, the source is rarely a concern unless the environmental professional were to question a finding that ragweed pollen has been identified, and there is no ragweed within a one mile radius of the sampling station. Pollen and spores do have the capacity to become dispersed beyond what would by the layman's terms be considered "highly improbable distances."

The amount of material discharged may be overwhelming. Fungal spores have been known to discharge as much as thirty billion spores per day. Pollen discharges can likewise be remarkable. Their numbers have been reported as high as seven trillion pollen grains per tree on a season. See Table 2.1.

Table 2.1 Pollen and Spore Single Source Discharge Rates

FUNGAL SPORES—one colony or growth unit	
Ganoderma applanatum	30 billion per day
Daldinia concentrica	100 million per day
Penicillium spp.	400 million per day
POLLEN GRAINS—one tree	
European Beech (*Fagus sylvaticus*)	409,000,000 per year
Sessile Oak (*Quercus petraea*)	654,400,000 per year
Spruce (*Picea abies*)	5,480,600,000 per year
Scotch Pine (*Pinus sylvestris*)	6,442,200,000 per year
Alder (*Alnus spp.*)	7,239,300,000 per year

Excerpted from *Sampling and Identifying Allergenic Pollens and Molds.* [2]

Attempts have been made to identify allergenic pollen types and the times of the year when their local presence is increased. Some highly allergenic individuals make decisions for relocation based on the prevalence of given allergens. Although Table 2.2 demonstrates an effort to categorize by state, the determinations are generalized and may not be representative of local areas within the regions mentioned.

The size, shape, and density of the airborne allergens affect their aerodynamic characteristics while the air humidity, wind direction, wind velocity, and obstructions affect their travel path as well as their distance. Temperature, soil types, and altitude may also impact the quantity of airborne allergens.

The size of fungal spores range from 1 to over 200 microns in diameter/length, although those which typify allergenic airborne spores range in size around 1 to 60 microns. The environmentally prevalent *Aspergillus* and *Penicillium* spore types are on the small end of the range, around 1 to 2 microns in diameter. It should be noted that some spore producing bacteria are also on the order of 1 micron in size and may appear microscopically to be mold spores. See Figure 2.2 for differentiation between two molds of similar spore production. The frequently identified spore types found on the larger end of the size scale are *Fusarium* (around 60 microns) and *Alternaria* (sometimes

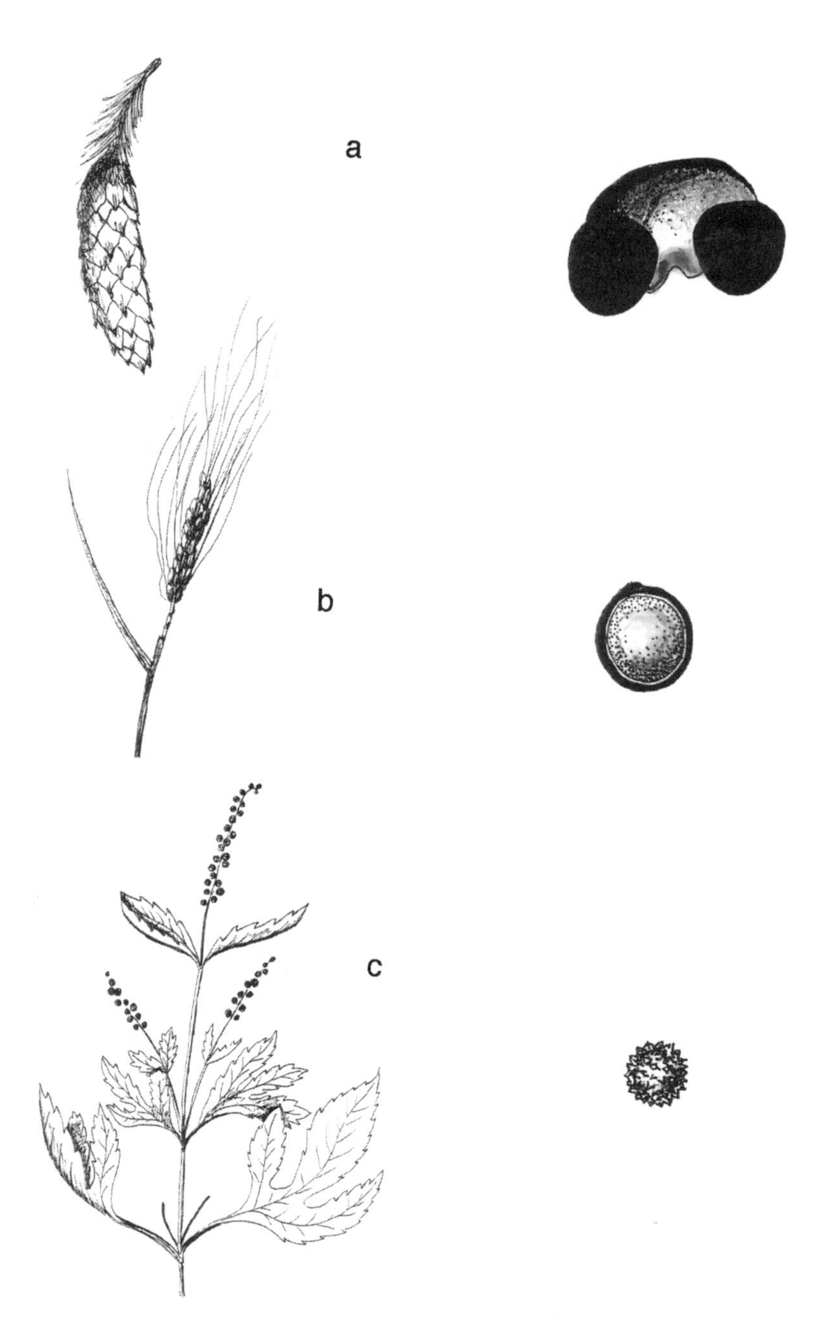

Figure 2.1 Representation of allergenic plants and pollen categorized into trees [e.g., cedar (a)], grasses [e.g., tall wheat (b)], and weeds [e.g., giant ragweed (c)]. The examples shown above are amongst the more allergenic within their own category.

up to 100 microns). The common outdoor *Cladosporium* varies from 4 to 20 microns). *Corynespora cassilicola* may reach lengths of 220 microns.

Pollen densities typically range from 19 to 1003 grains per microgram. Hickory pollen, which is moderate in size, are in on the low end. Giant ragweed and nettle, even though small in size, are on the high end in density. up to 100 microns). The common outdoor *Cladosporium* varies from 4 to 20 microns. *Corynespora cassilicola* may reach lengths of 220 microns.

Figure 2.2 Differentiation between molds starts with the microscopic appearance of colonies and spores. This example demonstrates the colony appearance of two different genera of fungi that have similar spores, both around 1 to 2 microns in diameter, which is also the some as that of the larger spore-forming bacteria. The above drawings are: (left) *Aspergillus spp.* and (right) *Penicillium spp.*

Variations in shape include spheres, ovals, spirals, elongated stellate (star-shaped), and clubbed. They may be elongate, chained, or compact, and, generally, the surfaces are smooth. They are lacking in hairs, spicules (needles), and ridges. See Figure 2.3 for some shape differentiating features.

The size of pollen grains shift to a larger size. They range from 14 microns (for stinging nettle) to 250 microns in diameter (for pumpkin pollen). Grass pollen are usually around 20 to 40 microns. Tree and weed pollen are the more variable. Most, however, fall between 20 and 60 microns. Red cedar and Western ragweed pollen are on the low end, around 20 to 30 microns. Scot's pine and Carolina hemlock are between 55 and 80 microns. Note that the specific tree and weed types are mentioned in reference to size. Unlike the fungal spores, the tree pollen are variable not only between genus but species as well.

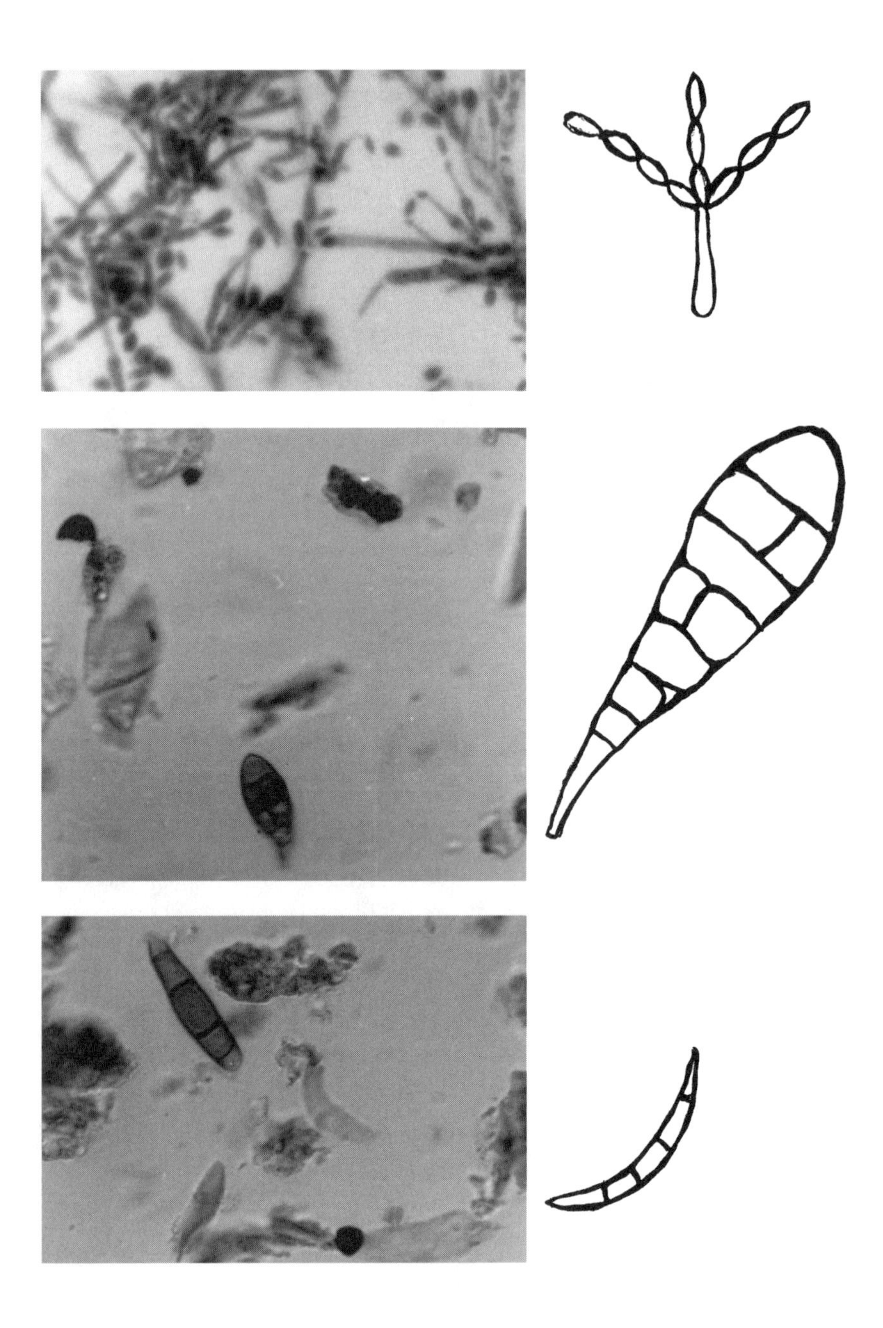

Figure 2.3 In addition to the fungal genera of *Aspergillus* and *Penicillium*, (top) *Cladosporium*, (middle) *Alternaria*, and (bottom) *Fusarium* are amongst the more commonly found mold spores. Spores stained with Lactophenol Aniline Blue Stain for best photoimaging contrast with a green filter. Photomicrographs under 400x.

Cedars vary from 20 microns (for Red Cedar) to 100 microns (for Atlantic Cedar). The grass and weed pollen do not demonstrate as extreme a variance.

Plant pollen are generally more complicated in design than are the spores. They tend to be spherical or elliptical with surface structures and/or pores, and the interior portions typically have a recognizable arrangement. They may be lobed with a smooth surface or spherical with spicules. Their interiors may be thick walled, undifferentiated or thin walled, multifaceted. Ragweed pollen have a spherical morphology with multiple spines. The pine pollen are lobed with a smooth surface.

Pollen densities range from 19 to 1003 grains per microgram. Hickory pollen, which are moderate in size, weigh in on the low end. Giant Ragweed and nettle, even though small in size, are on the high end in density.

STRATEGIC APPROACHES TO AIR SAMPLE COLLECTION

Intent must be clear! Armed with the reason for sampling and with the ultimate use of the results, the environmental professional can intelligently develop a strategy as to where and when sampling must be performed.

Historically, the concern is for outdoor air and its potential allergenic impact on the local populations. Limited studies have been performed to determine the variances between spore and pollen counts taken simultaneously at the same site (matched sampling) and comparing them to that of other locations. Not any have been published. Consequently, the environmental professional may seek to perform some local comparisons, initially and routinely, should the purpose be to provide information to the public. Consider the audience and their areas of concern. They may be in downtown Chicago, or they may be in the suburbs. Most people don't work outdoors, and the indoor office air may vary considerably from that outdoors. Yet, the spore and pollen count is of interest to motorists in transit and residents living in the city. Nonroutine sampling may, however, be indicated.

An allergist may request information regarding a specific site, or a office building manager may be concerned about occupant complaints. There may be an expressed concern regarding indoor air quality and allergy symptoms potentially associated with spores and pollen. The source site of the concern(s) must be considered. The environmental professional may choose to perform comparative sampling. If the indoor air has a predominant amount of ragweed, not found outdoors during the sampling period, the pollen may have previously entered the building and become entrained within the air handling system. An attempt to compare spore types, however, would not be possible, and should the counts appear excessive, the professional should consider viable allergen sampling techniques to obtain more information than the total count approach will allow.

More extensive evaluations may be indicated to identify the source by tracking concentrations. This may be planned following clarification as to the existence of elevated levels of a known allergen.

The audience must, once again, be considered, along with the required sampling duration for the device to be used for collecting the sample, when deciding on time. If the audience is residential suburbia, a daily sample may be desirable with a 24-hour sampling period. Weekly reporting may, however, be more feasible, in which case, some samplers may be programmed to take the sample over a 7-day period. If the audience wants office hour information yesterday, the sample should be taken during work hours with a minimum time involvement.

In brief, experience and a well thought out approach to sampling are a must. The where and when of sampling may also impact choices of methodologies and equipment.

Table 2.2 Plant Allergens by Region

NORTHERN WOODLAND
- Trees (April – June) — birch
- Fungi (June – October) — mushrooms and puffballs; watertight cabins and cottages tend to be moldy

EASTERN AGRICULTURAL
- Trees (March – May) — ash, birch, box elder, elm, mulberry , oak, sycamore, and walnut
- Grasses (May – July)
- Weeds (July – September) — hemp, goosefoot, and ragweed
- Fungi (May – November)
- Other — castor beans, cottonseed, and soybeans

SOUTHEASTERN COASTAL
- Trees (February –April) — ash, elm, oak, pecan, and sycamore
- Grass (February – October)
- Weeds (July – October) — ragweed
- Fungi (all year)

SOUTHERN FLORIDA
- Trees (January – April) — oak
- Grass (January – October)
- Weeds (June – October) — ragweed
- Fungi Indoors (all year)

GREAT PLAINS
- Trees (February – April) — oak
- Grass (April – September)
- Weeds (July – October) — goosefoot, ragweed, and sage
- Fungi (May – November)
- Other — livestock dander, fertilizer dust, animal feed dust, and grain storage dust

Table 2.2 (continued)

WESTERN MOUNTAIN
- Trees (January – March) — mountain cedar
- Grass (May – August)
- Weeds (July – October) — goosefoot, ragweed, and sage

GREAT BASIN
- Weeds (July – September) — goosefoot and sage

SOUTHWESTERN DESERT
- Trees (January – April) — Arizona cypress, mountain cedar, and mulberry
- Grass (March – October)
- Weeds (April – September) — goosefoot and ragweed
- Fungi (increased by use of evaporative cooling units in buildings)

CALIFORNIA LOWLAND
- Grass (March – October)

NORTHWEST COASTAL
- Grass (May – September)
- Weeds (May – August) — goosefoot

ALASKA
- Other — dog dander

HAWAII
- Grass (all year)
- Fungi Indoors and Outdoors (all year)

PUERTO RICO
- Grass (all year)
- Other — insect parts, bat droppings, and smoke of burning sugar cane (irritant or allergen unclear)

Excerpted from the U.S. Pollen Calendar.[3]

AIR SAMPLING METHODOLOGIES

Most outdoor sampling is performed to keep the public informed as to daily trends and to assist local allergists diagnose sources of bronchial asthma and hay fever. Presently, there are several hundred private enterprises which perform this service and report to their local newspapers routinely. Yet, there are less than one hundred certified reporting stations within the continental United States.

There is an attempt at the present by the National Allergy Bureau, a program under the American Academy of Allergy, Asthma, and Immunology (Milwaukee, Wisconsin), to certify/supervise readers and report collected airborne mold spore and pollen data from the various certified stations throughout the United States. The data is compiled within a manual on an annual basis and made available to the public for purchase. Should the reader desire to purchase a copy with region-specific information, be forewarned that not all states have recording stations and some states have several. The data is limited by the number of certified readers and where they are located.[4]

Although there are still those who collect samples, passively, on a greased microscope slide, sampling is generally performed using a centrifugal or suction air sampling device. The National Allergy Bureau recommends the use of the Rotorod or a wind oriented suction sampler. Of the latter, two samplers, approved for use by the National Allergy Bureau, are the Burkhard Samplers and the Kramer-Collins Spore Samplers.[4] Most of the sampling performed in the past has been through the use of the Rotorod, and adequate spore typing is only possible through culturing, an added feature of the Burkhard Sampler.

ANALYTICAL PROTOCOLS

Samples are analyzed by optical microscope. There are a limited number of commercial laboratories available to perform this service. Palynologist (associated within the Botany Department of some universities or in private practice) may assist the environmental professional. On occasion, microbiologists get involved in pollen identification as well. Then, there are university courses and infrequent one day extension courses.

Western Oregon State College (Division of Continuing Education and Summer Programs, Monmouth, Oregon) has a four day course on the sampling and identification of pollen and fungus spores, and Harriet Burge of Harvard University teaches a one day "certification" course on spore and pollen sampling/analysis at two different professional conferences each year. Several reference books are available to assist the professional in identifying molds and/or pollen. The most popular source information manual is Smith's Sampling and Identifying Allergenic Pollens and Molds (Blewstone Press, San Antonio, Texas).

Keep in mind that there are in excess of 350,000 different "known" plant species and many more that have yet to be classified. Not even the most experienced palynologist will be able to identify all species. Again, experience is the key. The more experience the reader has the better the identification competence. Then, too, many of the pollens are local. Whenever the analyst has an opportunity to collect a pollen sample from an identifiable local plant, he should do so and indicate known genus and species, if possible. A private library of microscope slides with local pollen, mounted and stained by the same analyst and staining technique to be used, is strongly advised. The pollen specialist (e.g., palynologist) will maintain his own library of reference samples for local pollens and those which have been accessible to the specialist.

The final results should be reported as total fungal spores and pollen grains (both viable and nonviable) per cubic meter of air. Oftentimes, a total count is provided, and only the pollen is identified. Some attempts, however, have been made by noncertified counters to identify fungal spores. Keep in mind, many mold spores of different genera appear microscopically similar to one another, to mushroom spores, to yeasts, and bacterial spores.

In those cases when attempts have been made to identify fungal spores by morphology, the results are precarious. Although some mold spores have well-defined structures, most require the vegetative structure and complicated culturing for proper identification. There are several genus types which are spherical in shape and approximate four to seven microns in diameter. Some of these are allergenic molds (e.g., *Aspergillus*). Some are nonallergenic molds (e.g., *Gliocladium*). Others are pathogenic molds (e.g., *Blastomyces*). Then, too, the spherical bodies may be yeast cells or mushroom spores. The author believes that attempts are made to identify molds, because the previously mentioned reference manual includes a section on fungal identification.

An optical microscope is required for counting and identification. The whipple disk is generally used to perform the count with a 40x objective. Identification, sometimes using the 100x oil immersion, is done by hydration, color staining. As there are different staining techniques, and the published photomicrographs have not consistently used for the same staining methods for identification of pollen and spores, the details for staining are intentionally omitted. Whatever stain is used, the professional must assure the spores and pollen grains are hydrated, stained in a fashion similar to that used for published or private library comparisons.

Attempts have been made by some to melt mixed-cellulose ester filters using hot, vaporized acetone. Not only does this not hydrate the allergens but dehydrates the already dry material. Prior to hydration, a spore/pollen grain typically lacks form and definition. Staining/hydration alters not only the appearance but size. Both are relevant for identification purposes. Staining and sample manipulation consistency is important when creating a library of slides. A desiccated, shriveled sample becomes unworkable!

INTERPRETATION OF RESULTS

According to the National Allergy Bureau, the mean threshold counts for allergy-like symptoms are subjectively in excess of 20 pollen grains per cubic meter of air (grains/m^3) for ragweed and for grass pollen. Although having a higher threshold for symptoms, tree pollen has not been assigned a numeric threshold. Mold spore thresholds are in excess of 1000 particles/m^3.

The National Allergy Bureau has further suggested definitions for site comparisons. See Table 2.3. The levels indicated are based on ecological measurements, not health effects. However, assuming dose/response relationships

Table 2.3 Comparison Values of Allergenic Plant Pollen and Mold Spores

	Low	Moderate	High	Very High
Trees	0-15	15-90	90-1500	>1500
Grasses	0-5	5-20	20-200	>200
Weeds	0-10	10-50	50-500	>500
Molds	0-900	900-2,500	2,500-25,000	>25,000

Excerpted from Parameters for Pollen/Spore Charts.[5]

are the same across the country, the thresholds are probably more appropriate than a limit, based on one single source.

The industrial hygiene professional organizations have typically monitored for viable microbials. The scenarios vary as to how to interpret viable counts. In 1985, the National Institute for Occupational Safety and Health recommended a limit of 1000 colony forming units per cubic meter of air (CFU/m^3) for viable microbes. Many would suggest this limit is too high.

The percent of allergens which are viable versus nonviable will vary considerably from one sample to the next, particularly wherein the sample sites are different. The nonviable microbes are just as allergenic as the viable ones. Thus, it is difficult, if not impossible, to evaluate viable sample results solely on a numeric threshold.

The most direct approach to interpreting outdoor allergenic pollen and spores is total counts of pollen grains with the type of plant identified and of spores. However, these thresholds should be used as guidelines only. Local differences will impact the final review process.

Indoor air evaluations may require some additional considerations. The counts should be compared to outdoor counts with pollen types identified and possibly to indoor noncomplaint areas. Professional judgment and experience are important in these instances. Use of viable allergen sampling methodologies may provide additional information particularly wherein fungal spores are suspected.

REFERENCES

1 Smith, E. Grant. *Sampling and Identifying Allergenic Pollens and Molds*. Blewstone Press, San Antonio, Texas (1990).

2 Ibid. p. 16.

3 Aeroallergen Network of the American Academy of Allergy, Asthma and Immunology (AAAAI). U.S. Pollen Calender. AAAAI, Milwaukee, Wisconsin, July 1993.

4 Bleimehl, Linda. Issues concerning the National Allergy Bureau. [Oral communication] AAAAI, Milwaukee, Wisconsin, January 1996.

5 Pinnas, Jacob L. Parameters for Pollen/Spore Charts. [Letter] National Allergy Bureau, Tucson, Arizona, November 28, 1994.

Chapter 3

VIABLE MICROBIAL ALLERGENS

Viable means "capable of growing and completing a life cycle." Understanding this is key to choosing appropriate sampling methodologies. Pollen types are identifiable by microscope irrespective of their viability. Yet, without using elaborate analytical techniques (which are discussed in another chapter), fungi, bacteria, rickettsia, and viruses can only be characterized when their viability is retained during the sampling process.

Keep in mind, however, that viability is not the driving force which causes allergic symptoms, and not all viable microbes are allergenic. Some are human pathogens (e.g., tuberculosis). Many affect our lifestyles (e.g., mildew in the home) and crops (e.g., crop mold). Others have no visible, direct impact on our lives. They may all become part of a sample and require differentiation from one another, based upon their potential impact on the population of concern.

Major outdoor allergens are plant pollen and fungal spores. Others include agricultural seed husks and algae. The seed husks dislodged into the air while managing grain elevators can be a source of considerable allergy problems to those in the immediate vicinity of a facility. For these situations, awareness as to the potential and likelihood of associated allergic problems is the simplest means for ascertaining that a problem exists. Algae is not recognized as an outdoor allergen.

The principal indoor allergens are amplified fungal spores and spore-forming bacteria. Others include house dust, mites, animal dander, and feathers. It should be noted also that indoor nonallergenic bacteria are generally amplified in office buildings as well. This is due to the proximity of the occupants and confinement of contaminated air. Another potential indoor allergen which is of minimal concern yet may be worthy of consideration is algae which will be addressed in this chapter.

Relevant to this chapter, the various airborne viable microbial allergens are described. The means for growth and amplification are discussed, and sampling methods are explained.

TYPES OF VIABLE ALLERGENIC MICROBES

Viable allergens are herein discussed to lend an understanding to the reader as to the potential sources and growth requirements needed for enhancement, amplification of allergenic microbes. The principal allergen in each case is thought to be airborne spores with minimal concern for the associated growth structures. The two major categories are fungi and thermophilic actinomycetes.

Those microbes which have been excluded from this allergenic microbes list are either pathogenic or have not been reported as probable allergens. The pathogenic microbes (represented by fungi, bacteria, viruses, and protozoa) are discussed separately in Chapter 4.

Fungi

Fungi, numbering over 100,000 different species, are neither plant nor animal. Lacking in chlorophyll (plant-like) and motility (an animal characteristic), they belong to a kingdom of their own. The Fungi Kingdom consists of molds, yeasts, and mushrooms. Where the yeast are single-celled organisms, molds grow into long, tangled strands of cells that multiply, forming visible colonies of varying sizes, shapes, and coloration. Mushrooms are tightly compacted masses of mold-like forms.

Molds

The cell wall and protective shell of molds are composed of polysaccharides (e.g., cellulose) and glucose units containing amino acids (e.g., chitin), both potential antigens. The reproductive process involves the release of thousands of allergenic spores, each having the ability to reproduce long, thread-like hyphae which continue to branch and form mycelium. The mycelia, in turn, attaches to a nutrient substrate and grows, unrestrained by size. As long as the mycelium has nutrient and room to grow, a single mycelium may theoretically expand to a diameter of fifty feet.

The protective cell walls and associated spores, material and antigenic components of the various species differ in complexity and allergy causing capabilities. Although the spores are generally implicated in allergenic conditions due to sheer numbers and ease of dispersion into the air, sections of the growth structures can be equally allergenic.

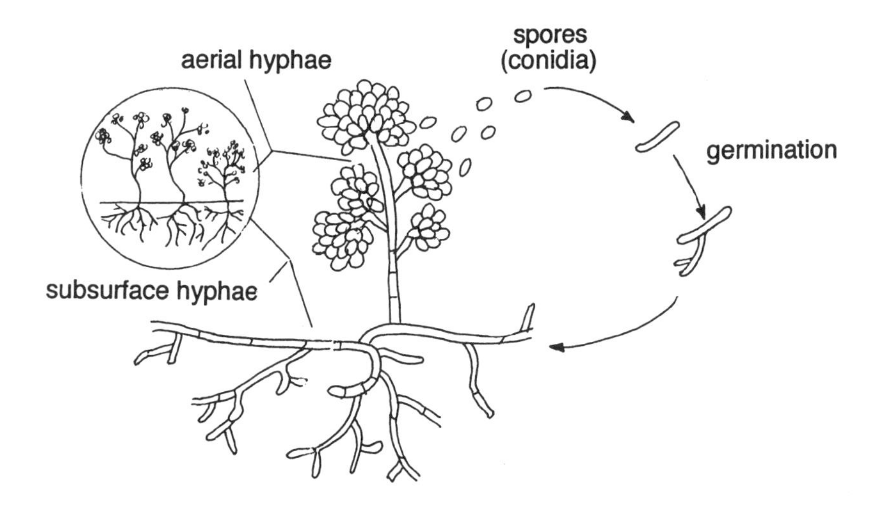

Figure 3.1 Typical Mold Structures

Specific mold genera are reputed to provoke allergy-like symptoms more consistently than others. Although there are varying opinions as to which one of these are allergenic, the nonallergenic molds may also contribute to the overall response of an individual's immune system. See Table 3.1. Some authorities believe that individuals can develop an allergy to a nonallergenic fungi and be genus specific to a fungal spore which is not commonly a problem for most people. Generally, however, allergenicity is genus specific.

Table 3.1 Mold Spores Reported to Provoke Allergy Symptoms

Acremonium spp.[1]	*Nigrospora spp.*[3]
Alternaria spp.[1, 2, 3, & 4]	*Penicillium spp.*[1, 2, 3, & 4]
Aspergillus spp.[1, 2, 3, & 4]	*Phoma spp.*[3]
Aureobasidium spp.[3]	*Rhizopus spp.*[3]
Aureobasidium pullulans[1]	*Scopulariopsis*[3]
Chaetomium spp.[3]	*Stachybotrys spp.*[1]
Cladosporium spp.[3]	rust molds and smuts[4]
Drechslera spp.[1 & 3]	
Epicoccum spp.[3]	
Fusarium spp.[3]	
Helminthosporium spp.[4]	
Mucor spp.[3]	

The most common airborne spore is *Cladosporium.* Beyond *Cladosporium*, there is some variation, based on geographic region and the time of the year. The consensus appears to be for *Alternaria* as the second largest contributor, and many include *Aspergillus* and *Penicillium*. Ironically, all these molds are reputed causative allergenic agents for most mold-sensitive patients. Table 3.2 provides percents of total airborne fungal spore information, as disclosed by one source and a summary of known natural habitats. Most findings include many of the same genera with a slight variation on percents, based on regional differences.

Contrasted with the hardiest of microbes, molds not only can stay alive indefinitely on inanimate objects, or "fomites," but grow into and destroy wood, cloth, fabrics, leather, twine, electrical insulation, and many other commercial products. They destroy lenses of microscopes, binoculars, and cameras. In localities where humidity is high, fungi do great harm to wood structures, telephone poles, railroad ties, and fence posts. Most of these problems are reduced by means of artificial preservatives. Sometimes the trouble starts in forest lands where fungi invade the heartwood and cause wood rot before the timber has had a chance to be cut down. The humid Amazon rain forest is one such example.

Thousands of products are treated to prevent decay, yet there are some types of fungus that thrive on preservative-treated wood. One such example is creosote-treated railroad ties! Other fungi-specific nutrients are vinyl wall covering adhesives, gypsum board, cellulose-based ceiling tiles, dirt retained within carpeting, and surface paints. Some feed on plywood. Others consume the glue used to laminate wood which is used in airplanes, furniture, and cars and will cause the layers to separate. Books and leather shoes are readily consumed by the microbes which are visible as mildew. Aircraft electrical systems, operating in tropical climates, require protection against insulation consuming molds. Immune suppressed individuals (e.g., AIDS patients) can also be host to nonpathogenic, invading molds.

Exterior molds grow on decayed organic material, corn, wheat, barley, soybeans, cottonseed, flax, and sun-dried fruits. They consume other plants, vegetable matter, and decayed organic material (e.g., dead animals).

Pathogenic molds parasitize and obtain nutrients from a host. The host may be plant or animal, and these molds vary slightly from the nonpathogenic molds both in environmental and sampling requirements. Thus, they are discussed in greater detail in Chapter 4.

Most molds require high moisture content. The genus *Stachybotrys* colonize on wet ceiling tiles. Some do quite well in minimal moisture, such as the "zeophylic (dry-loving)" fungi. *Aspergillus versicolor* prefers slightly moist gypsum board. Water requirements for an abbreviated list of microorganisms are presented for comparative reference in Table 3.3.

Table 3.2 Most Common Airborne Allergenic Fungi Genera from Nineteen Random Surveys

Genera	Prevalence (Percent)	Natural Habitats
Cladosporium	29.2	Worldwide: soil, textiles, foodstuffs, and stored crops
		Region dependent: woody plants (straw), and paints
Alternaria	14.0	Worldwide: decaying plant matter, foodstuffs, soil, and textiles
Penicillium	8.8	Region dependent: soil, decaying vegetation, foods, cereals, textiles, and paints
		Occasional occurrences: composts, animal feces, paper/paper pulp, stored cold temperature foods, cheese, and rye bread
Aspergillus	6.1	Region dependent: soil, stored cereal products, soil, food stuffs, dairy products, textiles, composts, and house dust
		Occasional occurrences: spoiled fruits/vegetables, stored grain/straw/cotton, meat, wool, dairy products, and salted foods
Fusarium	5.6	Worldwide: soil and plants
Aureobasidium	4.7	Worldwide: soil, decaying pears and oranges, paint, wood, and paper

Excerpted from *Mould Allergy*.[4]

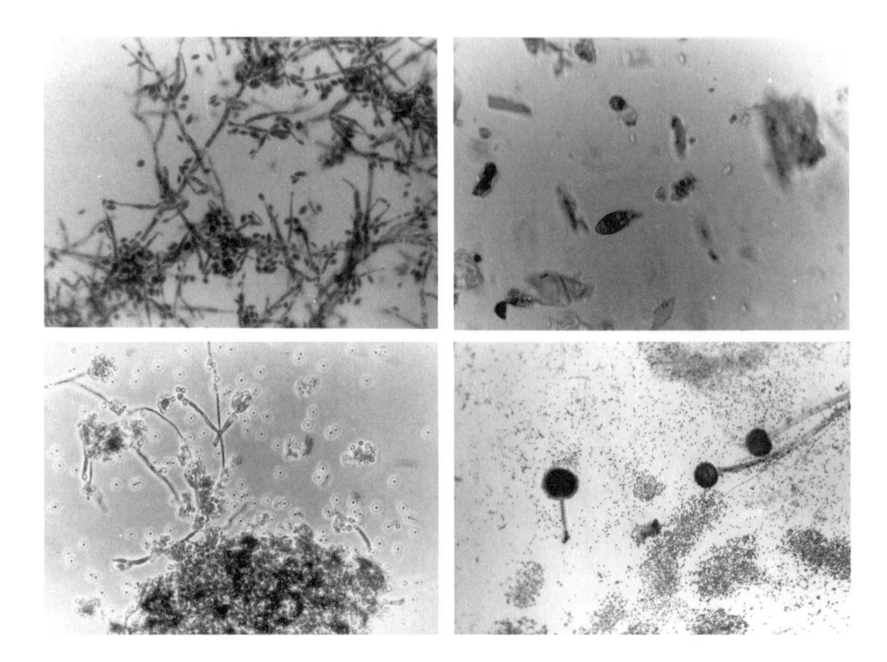

Figure 3.2 Photomicrographs of most common allergenic fungi and their spores. These are: (top left) *Cladosporium spp.*, (top right) *Alternaria spp.*, (lower left) *Penicillium spp.*, and (lower right) *Aspergillus niger*, all stained with Lactophenol Aniline Blue Stain and magnified 400x.

Table 3.3 Moisture Relations of Storage Microorganisms

Microorganism	Water Activity
Aspergillus halophilicus and *Aspergillus restictus*	0.65-0.70
Aspergillus glaucus and *Sporendonema sebi*	0.70-0.75
Aspergillus chevalieri, Aspergillus candidus, Aspergillus ochraceus, Aspergillus versicolor, and *Aspergillus nidulans*	0.75-0.80
Aspergillus flavus, Aspergillus versicolor, Penicillium citreoviride, and *Penicillium citrinum*	0.80-0.85
Aspergillus oryzae, Aspergillus fumigatus, Aspergillus niger, Penicillium notatum, Penicillium islandicum, and *Penicillium urticae*	0.85-0.90
Yeasts, bacteria, and many molds	0.95-1.00

Excerpted from *Microbiological Ecology of Foods.*[5]

Temperature preferences are variable as well. Although most molds do well at room temperature, some flourish at elevated temperatures. Others have an affinity for cold climates. The ranges are presented in Table 3.4. Yet, most spores favor moderate temperatures.

Table 3.4 Temperature Relations of Common Mold Species

Temperature to kill most "mold spores" (within 30 minutes)	140-145° F
Maximum growth temperatures	86-104° F
Optimum growth temperature	72-90° F
Minimum growth temperature	41-50° F

Excerpted from *High Protista.*[6]

Fungi tend to grow more during months when the humidity and temperature are elevated. In some regions, the peak mold spore season is in the spring, followed by the summer months. Other areas of the country experience peak periods in the fall. The winter months typically provide the least accommodating conditions for fungal growth. Although the daily and monthly variability is based entirely on humidity and temperature, growth will increase or decrease at certain hours of the day or night regardless of the outdoor climate.

Many, not all, fungi have peak growth times which are genus, sometimes species, dependent. Some peak in the late at night (e.g., *Cladosporium* and *Epicoccum*). Others peak in the early morning. Some peak in the late afternoon (e.g., *Alternaria* and *Penicillium*). A few peak, irrespective of time frame, immediately after a heavy rainfall.[4] Studies vary on their opinion as to these times, yet they all agree that peak periods do exist.

Indoor air environments may vary in humidity content of the air along with the outdoor air environment, and peak mold seasons may correlate to the indoor environment, particularly where humidity controls are not maintained. High humidity (greater than 60% relative humidity). is the leading causes of fungal amplification within buildings. Other means of mold spore amplification include, but are not limited to: (1) settled water sources (e.g., air handling system drip pans); (2) damp building materials (e.g., wet ceiling tiles); (3) air movement from a hot, humid crawl space into an occupied office area; (4) disturbances of settled dust (e.g., dry dusting); and (5) poor vacuum cleaner filtration. It should also be noted that mushrooms have been identified in buildings where water-damaged carpeting has been left uncorrected and other areas where there is an accumulation of water (e.g., behind leaking washing machines).

Molds grow in wide ranges of pH. Although pH 5 to 6 is favored by most, some are found proliferating between pH 2.2 and 9.6. Rare forms have been found consuming nutrient impurities found in bottles of sulfuric acid.

As fungi typically require oxygen, they tend to grow in oxygen rich environments (e.g., air handling systems). Some, however, grow quite well in enclosed areas where the oxygen may be minimal (e.g., between vinyl wall coverings and the wall). Yet, they all do require some oxygen. There are no anaerobic fungi.

The reason for their persistence and survival competence is attributable to their wide range of habitats and their means of spore dissemination. One colony is capable of dispersing millions of spores a day. See Table 3.5. The spores are shot out of their capsule or dislodged from their stalk and carried by the wind to be spread far and wide. Spores (and pollen) travel, in extreme cases, as far as 1,500 miles,[7] and it is common to find them a hundred miles from their point of origin. More simply stated, their source does not necessarily have to be in the immediate vicinity.

Table 3.5 Number of Spore Discharges from One Source

Ganoderma applanatum	30 billion/day
Daldinia concentrica	100 million/day
Penicillium spp.	400 million/day

Excerpted from *Sampling and Identifying Allergenic Molds*.[8]

Yeasts

Yeasts, one-celled fungi, are usually spherical, oval, or cylindrical in shape. They usually do not form filamentous hyphae or mycelium. A "population" of yeast cells remains a collection of single cells. They can be differentiated from bacteria only in their size and internal morphology (with an obvious presence of internal cell structures). Some of the yeasts reproduce sexually. The sexual reproductive process forms ascospores which are (like mold spores) resistant to many environmental conditions.

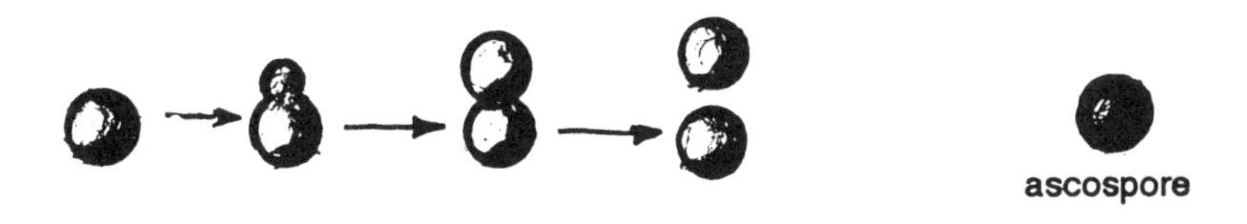

Figure 3.3 Common Yeast and Ascospore

Yeasts usually flourish in habitats where sugars are present (e.g., fruits, flowers, and the bark of tree). The most important ones are the baker's and beer brewer's yeasts. These have been selected and manipulated by man. They do not serve as good representatives of the classification. They are atypical.

Single-celled rust and smut proliferate to form thick-walled, binucleate spores. With an excess of 20,000 species, "rust" fungi are so referenced due to an orange-red color imparted to diseased plants by their urediospores.

"Smut" fungi have over 1,000 . The term smut is assigned to this class of fungi, because teliospores impart a black, sooty appearance to plants. Those which are routinely found in surveys of airborne fungi are *Cryptococcus* and *Rhodotorula*. The levels of smut indoors generally equals that of the outdoor air.[8] It is thus improbable that amplification occurs in office buildings.

Mushrooms

Mushrooms are filamentous fungi that typically form large structures, called fruiting bodies. These fruiting bodies are commonly referred to as the mushroom cap. The reproductive mechanism is completed within this cap, and spores are discharged into the air to contribute to the total airborne fungal spore numbers. It is unlikely that growth and amplification of the mushroom spores occur indoors without intent (e.g., cultivation of edible mushrooms).

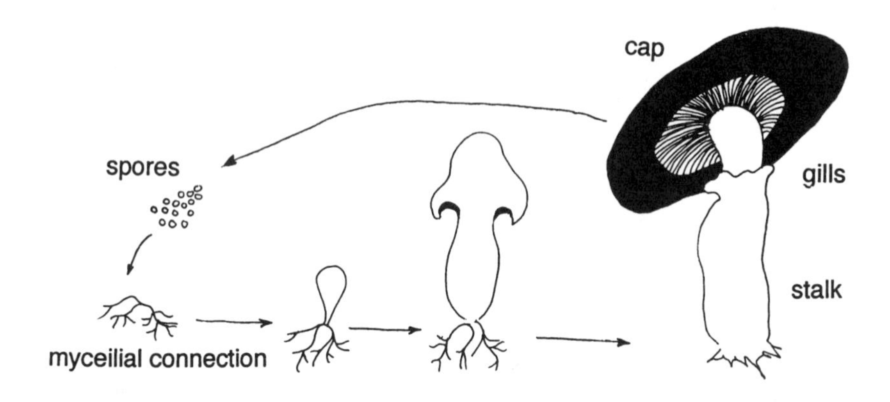

Figure 3.4 Life Cycle of Mushrooms

Slime Molds

Slime molds are not true fungi, because they lack, for most of their lives, a cell wall. The term "slime mold" refers to their swarming bodies of amoeboid cells during part of their life cycle. In this stage, many of the slime molds display brilliant colors, appear mucoid on a nutrient surface.

They live primarily on decaying plant matter (e.g., leaf litter and logs) and bacteria-rich soils. Their food consists mainly of other microorganisms, especially bacteria, which they ingest by phagocytosis. In this stage of their life, they do not pose a problem.

The reproductive cycle, however, involves the formation of stalks which produce spores (or multiple spore-containing sporangia). Although they do produce spores in small numbers, slime molds can be contributors to the total fungal spore count. It is unlikely that growth and amplification occur indoors. Thus, amplification indoors is not generally a concern.

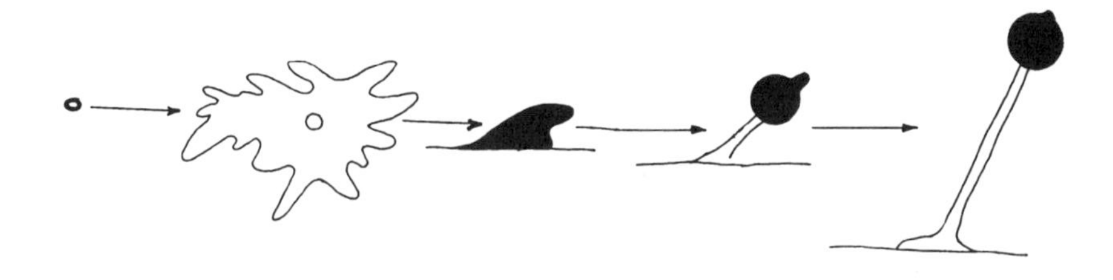

Figure 3.5 Life Cycle of Slime Molds

Thermophilic Actinomycetes[10]

Actinomycetes are filamentous bacteria that resemble fungi in their colonial morphology and production of spores. Under the microscope, colonial masses appear as thin hyphae (generally much thinner than those found in fungi) with associated spores which are at the low size range for fungal spores. They typically range from 0.8 to 3 microns in diameter. These may tend to be overlooked if the microscopist is performing a pollen and spore count only (without the benefit of culturing). They are spherical or oval in shape and have a preference for elevated temperatures.

Ideal temperature preferences range from 37 to 60° C. Yet, these ranges may be, at times, narrow and highly selective. *Thermoactinomyces candidus* grows rapidly between 55 and 60° C but will not grow at 37° C or less. Although some will grow at 37° C, most prefer 45 to 55° C.

Their nutritional requirements are complex, and their spores are more temperature resistant than most fungal spores. Thermophilic actinomycetes are the only recognized bacteria that form allergenic spores.

The most important genera are the *Micropolyspora*, *Thermoactinomyces*, and *Saccharomonospora*. Some species of *Streptomyces* have been implicated as well. In nature, these organisms generally require a nutritionally rich substrate and elevated temperatures. Ideal habitats include moldy hay, compost, and other vegetable matter. Indoor amplification may occur in the heating and humidifying systems where there is also a source of nutrients (e.g., vegetable matter build-up in an air handler with elevated temperatures).

AIR SAMPLING METHODOLOGIES

There are as many approaches to sampling for viable microbes as there are types of bioaerosol samplers. Some methods may be better than others wherein the intent and information sought in the various circumstances are highly variable. There are no governmentally controlled limits or requirements for monitoring. There have been a few attempts by professional organizations, universities, and private firms to provide guidelines. The most readily accepted is that set forth by a committee designated by the American Conference of Governmental Industrial Hygienists (ACGIH).

Recommended by ACGIH [11,12]

In 1986, the ACGIH published a report with the intent to provide a guideline to industrial hygienists for evaluating "workplace-related illness such as hypersensitivity pneumonitis, humidifier fever, and allergies that are likely due to bioaerosols."[13] They later published a more expansive variation with considerable changes in interpretation. This more recent guideline, dated 1989, is available for purchase through the ACGIH which is located in Cincinnati, Ohio.

Its principal purpose is to identify the "source of bioaerosol components so that effective corrective action may be undertaken."[14] Only where allergenic problems or illness indicate probable airborne microbes should the arduous task of air monitoring be conducted. They do not recommend routine sampling for reasons other than research.

Sampling Strategy

The sampling strategy is subject to the environmental professional's evaluation of a specific situation. Careful thought and planning are paramount!

A minimum sampling effort for fungal spores should include samples taken: (1) outdoors (preferably in the vicinity of the air intake for the indoor air for the area in question); (2) indoors prior to potential source disturbances; and (3) indoors following potential source disturbances.

They further suggest monitoring (for fungal spores and bacteria such as thermophilic actinomycetes) be performed in the vicinity of the people experiencing symptoms, a nonsymptomatic area, and outdoors. Plans should also consider monitoring potential source amplification areas (e.g., inside an air handling unit) for diagnostic reasons.

If the environment smells moldy, has visible signs of fungal or bacterial growth (slime and foam), the reservoir, once located, should be remediated. The ACGIH recommends bulk sample collection of the suspect source for identification later in a laboratory and possible confirmation of elevated airborne levels associated with the identified source. The sample should be collected in a clean (or sterile) plastic or glass container.

Equipment

Although there are several possible choices for sampling equipment, selection is situation dependent. However, in most cases, the air sampler(s) of choice for allergenic spores are one of the following two types:

- Culture and particulate impactors with known particle collection efficiencies.
- High-volume filtration sampling (where the antigen is known).

The ACGIH discussed the advantages and some disadvantages of each method. They are as follows:

Slit Impactor—The slit impaction sampler works either as culture or particulate impactor. It allows for sampling durations which provide for a greater chance that a disturbance or peak levels will occur during the sampling period. Plates also can be subdivided and analyzed in time sectors. The 24-hour and 7-day particulate samples provide time discrimination in units which may be abbreviated to one hour. In brief, the slit impactor, although not time tested, appears to provide fairly reliable information for viable and nonviable microbes while providing some time-differential information.

Sieve Impactor (e.g., Andersen impactor)—Accuracy of the Andersen sampler for viable spore sampling has been time-tested. The samples are time integrated over a limited sample duration. If a peak time is identified, this may allow for enumeration of brief peaks. The two-stage impactor is compared with the single-stage and found to be of limited value when sampling for allergens. The reason for this is that the separation of respirable and nonrespirable allergens lends no additional information than a total composite. Single-stage portable impactors are easier to manage, yet their sampling duration and efficiency are in question. The sample duration is usually less than one minute. In brief, the old time-tested sieve impactor provides an accurate means of sampling viable allergens only, over an abbreviated time frame.

High-volume Filtration—Familiarity and simplicity are the principal advantages to the filter cassettes. Although efficient for collecting down to filter pore-sized particles, the method of filter clearance (which has a severe drying effect on the collected allergens) and subsequent analysis frequently underestimates spore levels. Washing and culturing a smooth-surfaced filter (e.g., polycarbonate) collectate is less damaging and allows spore identification of the viable spores which were not harmed during the sampling process. The sampling duration for the high-volume sample is variable with a sample collection of up to 1000 liters. This method is not recommended for viable bacteria as damage is imminent. In brief, filter cassette sampling is limited to spores requiring composite time averages with no identification requirements necessary.

Air sampling for viable allergens generally provides an underestimation of the airborne particle loads, because allergenicity is due to viable and nonviable microbes, not to viable allergens only. Additional particulate sampling is recommended when sampling for allergens.

Culture Sampling Media

The recommended sampling media for airborne fungi is malt extract agar. It should support growth of most viable fungal spores and is considered an excellent diagnostic media for determining species of *Aspergillus* which may be important, not only for allergenicity, but diagnostic of species dependent illnesses. Other media alter fungal morphology, making visual identification difficult, and inhibit some of the fungal growths. Bactericides (e.g., Rose Bengal) may be added to the malt extract agar to keep the bacterial contamination to a minimum. The latter becomes a must only in environments where high bacterial levels are anticipated (e.g., agricultural environments).

Table 3.6 Samplers Commonly Used for Bioaerosol Sampling

Type of Sampler	Sampling Rate (liters per minute)	Recommended Sampling Time
CULTURE IMPACTION		
slit impactor 30 to 70	1 hour (continuous) 7 days	24 hours
sieve impactor		
single-stage impactor	28	1 minute
two-stage impactor	28	1 to 5 minutes
single-stage portable	90 or 185	0.5 or 0.3 minutes
PARTICLE FILTRATION		
filter cassettes	1 to 2	15 to 60 minutes or 8 hours
high-volume filtration	140 to 1400	5 minutes to 24 hours

Excerpted from *1989 ACGIH Guidelines—Air Sampling.*[15]

Samples should be transported to the laboratory within 24 hours, and the source sample processed and incubated immediately. Suggested fungal incubation is five days at room temperature with subsequent identification by genera. Counts should be made of each readily identifiable colony and results given in colony forming units per cubic meter of air (CFU/m^3).

There is no recommended sampling media for thermophilic actinomycetes. Yet, they require temperatures in excess of 50° C. The growth period is not well defined, and there is no guidance as to identification and speciation of the actinomycetes.

Recommended by a Commercial Microbiological Laboratory[16]

A prominent commercial microbial laboratory publishes a technical bulletin which details two microbial air sampling protocols which include the minimal 1989 ACGIH Guidelines. These protocols are intended to address indoor air quality concerns involving viable allergens and the potential presence of other microbes.

Sampling Strategy

They recommend sampling multiple samples at each sample location within the same sampling time period. They further suggest sampling be performed at each sample location at various times during the day (e.g., morning and afternoon or at 2 hour intervals). The ultimate decisions must be assumed, however, by the on-site investigator.

The two protocols are referred to as the four- and seven-plate sampling protocols and are summarized in Table 3.7. The total number of samples to be analyzed by the laboratory is considerably more than that of the ACGIH approach which involves two to three plates at each location. Once again, the ACGIH plates include commensal/environmental bacteria, thermophilic actinomycetes, and fungi.

While the four-plate protocol is intended to enumerate fungi and environmental bacteria, the seven-plate protocol has an added provision for enumerating human commensal bacteria as well as the environmental bacteria. Human commensal bacteria are amplified in densely populated areas. Although this approach allows for more targeted separation of environmental and human commensal bacteria, it does not provide a clean separation. Some environmental bacteria will grow in the same media, at the same temperatures as the commensal bacteria, and some commensal bacteria will grow in the same media, at the same temperatures as the environmental bacteria. Genus identification must be performed for type confirmation.

Equipment

The recommended sampling equipment is the N6-Andersen impactor. In all cases, the air flow rates are 28.3 liters, yet two sample durations are suggested. These are typically 60 seconds and 120 second sampling times, excepting a recommended 180 second sampling duration for thermophilic bacteria. The rationale is more thorough enumeration if the counts are low or marginal and simultaneous duplication of samples for quality control. Duplication is another divergence from the ACGIH protocol.

Table 3.7 Four- and Seven-plate Sampling Protocol

Plate	Sampling Media	Sampling Time	Incubation Period	Group Enhanced
1	RBA	60 seconds	RT	fungi
2	R2Ac	60 seconds	RT	environmental bacteria
3	BA	60 seconds	35C	human commensal bacteria
4	RBA	120 seconds	RT	fungi
5	R2Ac	120 seconds	RT	environmental bacteria
6	BA	120 seconds	35C	human commensal bacteria
7	TSA	180 seconds	56C	thermophilic bacteria

The Four-plate Sampling Protocol is indicated in the gray area. Room temperature (RT) is $23 \pm 3^{\circ}$ C.

Culture Sampling Media

The growth nutrients are summarized as follows:

- Rose Bengal agar (RBA) is used for isolating fungi while suppressing bacterial growth. An alternative which does not suppress bacterial growth is malt extract agar.

- R2Ac agar with cycloheximide (R2Ac) is a low nutrient medium designed to enhance the growth of environmental bacteria, but human commensal bacteria may grow as well. The cycloheximide suppresses fungal growths.

- Heart infusion blood agar (BA) which contains 5 percent sheep blood enhances growth of human commensal bacteria. An alternative media is tryptic soy agar. Elevated incubation temperatures assist in partial suppression of the environmental bacterial growths.

- Tryptic soy agar (TSA) is a high nutrient medium on which only the thermophilic bacteria will grow at the elevated incubation temperatures.

Comments and Considerations

Although it is presently rejected as obsolete, the 1986 ACGIH Guidelines discussed the topic of bioaerosol sampling more in depth than the more recent 1989 Guidelines. These details and a discussion thereof may contribute "food for thought." This information is provided for consideration only. There are no commonly accepted protocols for sampling.

The 1986 Guideline sampler of choice was the single-stage N6-Andersen impactor, with collection onto sterile Petri dishes with nutrient media. Other samplers were, however, recognized (e.g., Spiral Air, Biotest, Microban, and slit samplers). All these samplers and many more are discussed in the Appendices. When these other samplers are used, adjustment factors based on relative efficiency to the Anderson impactor were recommended (e.g., a sample that is 50 percent as efficient as the Andersen, multiply by 0.5 to adjust the results). This sometimes complicates the issue, especially where there are several different published efficiency comparisons.

For the N6-Andersen sampler, the directions were spelled out. Use a sterile plastic Petri dish (100 mm x 15 mm) and add 45 milliliter of media, or use a sterile glass Petri dish and add 27 milliliter of nutrient media. The size of the Petri dish may vary, dependent upon the specific Andersen impactor intended for use. Therefore, the amount of media poured may vary as well. If your Petri dish supplier does not know the size sample dish needed for specific samplers, Graseby Andersen (Smyrna, Georgia) will be able to provide this information.

The nutrient media for general detection and enumeration of viable fungi was malt extract agar. For bacteria (e.g., thermophilic actinomycetes), the media of choice was trypticase soy agar.

Nutrient media can be purchased, already plated on the Petri dishes, from a laboratory supply retailer (e.g., Remel or Difco), or a premixed malt extract powder may be purchased, with the instructions provided by the supplier. Although it is easier and preferable for the environmental professional to purchase the already plated Petri dishes, it may take as much as ten days to receive a rush supply of pre-plated Petri dishes. The manufacturer batch mixes only after receiving several orders. This time delay could be crucial. To confuse the issue, if the pre-plated Petri dishes are purchased and stocked by the environmental professional, the dishes dry out and can no longer be relied upon after thirty days in refrigeration with the protective cover sealed. A limited number of microbial laboratories do, however, assist in avoiding such problems or the potential problems associated with homemade sampling media.

Even though most of the laboratories leave the environmental professionals to seek out their own Petri dishes, the laboratory who is going to perform the final analysis should be consulted during the planning stages and prior to purchasing/preparing the sampling media. The laboratories deviate from each other in preferred sampling media. Colony identification is easier by their growth characteristics which are media dependent and media variable.

The N6-Andersen must be calibrated to a flow rate of 28.3 liters per minute with a one-minute sample duration. Where the anticipated results exceed 500 colony forming units per cubic meter (CFU/m^3), a larger air volume will result in excessive colonies which may obscure the final count, rendering the results unreadable.

WARNING: ***Do not attempt to disinfect an Andersen sampler with hydrogen peroxide. It reacts with and damages the unit while closing the sieve holes.***

The sampler(s) should be disinfected between sample locations. The ACGIH recommends immersion of sieve plate 70 percent ethanol for one minute. The excess should be drained off and the plate allowed to dry prior to its next use, and the sieve holes must be inspected prior to proceeding. The author's findings have been that this is impractical. The dry times and clearance of the sieve holes may take in excess of 15-minutes, up to 30-minutes. Another technique for assuring the samples don't get cross-contaminated has been to vacuum the sieve plates and interior of the unit with the vacuum pump, then allowing the sampler to run at the new location, "prior to placement of the Petri dish," for a couple minutes. The latter method may be used in conjunction with the 70 percent ethanol to speed up the clearance time for the sieve holes and assure the next sample doesn't get disinfected during the sample collection time.

Care should also be taken with the Petri dish cover for the duration of the sampling. A minimum precaution should be to place the cover face down on a clean, smooth surface for the duration of the sampling. If using an impactor and the nutrient media does not show regular impressions (see Figure 3.6), one should be suspicious that the seals were not seated properly or some of the holes were obscured. Where this occurs, the cause of the problem should be rectified and another sample taken.

After the sample has been taken, replace the cover, tape the edges, label the Petri dish, and store it with the agar side up. This assists in avoiding evaporative buildup with subsequent spreading of the impacted spores and their new growths.

The laboratory chosen to perform the sample analyses should be consulted prior to each scheduled sample collection for instructions as to their in-house procedures for packaging, shipping, and receiving. If shipment to a laboratory is required, the laboratory will probably require overnight shipping and, if possible, on the same day they were collected. Shipments could arrive on a Saturday or Sunday, requiring proper receipt coordination by the laboratory, or the laboratory may prefer receipt of samples to conclude with analysis on a weekday. Overnight shipping is necessary to insure that the plates arrive at the laboratory before any visible growth develops on the plates. This is also necessary because the human commensal bacteria and thermophilic bacteria must be incubated at elevated temperatures, and the environmental bacteria may be encouraged to grow at the temperatures encountered during shipping.

According to the old guideline, fungi should be incubated for three to seven days at 25° C with twelve hours of ultraviolet illumination per twenty-four hour time period. Fungi demonstrate highly variable growth rates. Some colonies become visible within 24-hours. Most are visible after three days, and rarely will they take as much as two or three weeks. For this reason, an attempt is made to get all existing, viable spores to grow so they can be seen by the naked eye while not allowing the faster growing fungi to overrun the plate. Identification of pathogenic fungi should not be counted on. They oftentimes require much longer growth times, up to three weeks. After three weeks, the other fungi will likely have already overgrown the Petri dish, and newly formed colonies would not be observable. The recommended incubation for thermophilic actinomycetes is 55° C for three days.

There have been reputed matched samples taken and analyzed by separate laboratories for comparison. When the results differ by an order of magnitude, the report differences are not surprising. Each laboratory has different incubation times and means for reporting their findings. Some report after three days, some after five days, and others after one week. There have been occasional reports provided by a laboratory made within two days of incubation. Early reporting is sometimes encouraged by the environmental professional and the laboratory attempts to oblige their needs. See Figure 3.6 for Petri dish growth samples.

Along with the colony count (reported in colony forming units per cubic meter), some laboratories also attempt identification of the more predominant genera, and most differentiate between fungal and bacterial counts. Additionally, the lab may provide different descriptors which will require further understanding by the environmental professional. Some laboratory-dependent descriptive terms follow:

Unknown—A colony that could not be identified. This may be due to any number of reasons which are laboratory dependent. Unknowns can occasionally be identified by additional, oftentimes extensive laboratory manipulation of the specimen.

Non Spore-Forming Body—Some colonies cannot complete their life cycle on a given nutrient. Without sporolation, identification becomes impossible. Such non spore-forming growths may be due to any of a number of microbes, some of which may be allergenic (e.g., yeasts, mushrooms, molds, smuts, and/or rust). The only alternative is the transference of the unidentifiable growth to other growth nutrients. Should you require this information, the laboratory time-expenditure will increase as will the fees.

Artifact—Some laboratories refer to a nutrient-estranged growth as an artifact. This may be a growth which occurs at a location other than the visible impaction depressions. An artifact may be along the sides, between the depressions, or disassociated with the nutrient.

Too Numerous to Count—A plate can become so overgrown with growths that a count may become obscured, impossible to distinguish one colony from the next. These situations may result from excessive air volumes.

Reporting terms are not standardized. A different term may be substituted for one of the preceding ones, or the definition may differ. Seek clarification by the laboratory which performs the analysis.

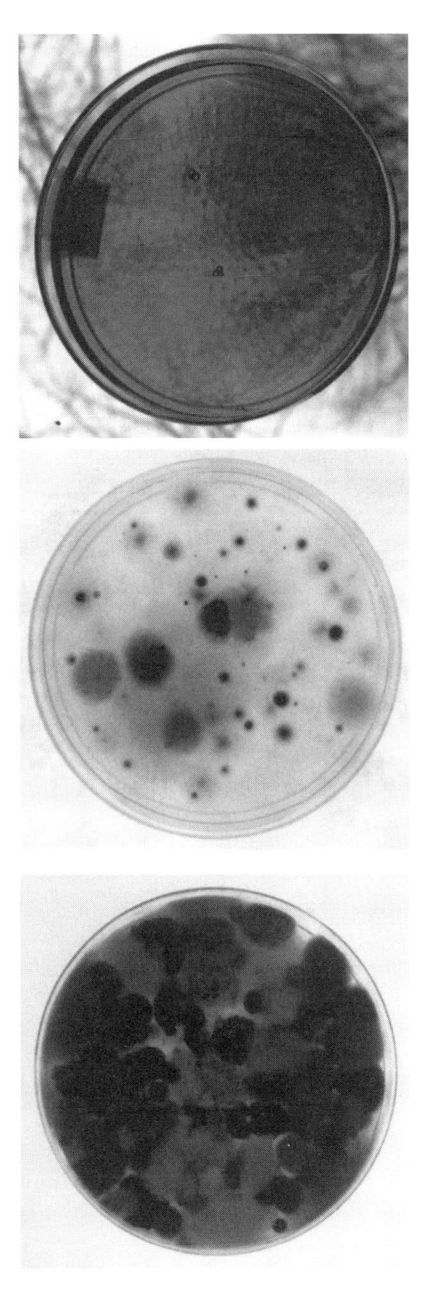

Figure 3.6 Petri dish deposition using impaction samplers. Examples are: (top) impaction marks on agar, (middle) culture with count of 1761 CFU/minute on malt extract agar after 5 days of incubation, and (bottom) excessive growth on malt extract agar after 5 days of incubation.

INTERPRETATION OF RESULTS

The interpretation of results is "highly" variable and controversial. Ultimately, the environmental professionals decide on the appropriate interpretation for a given situation. Attempts have been made by various researchers and professional groups to set an exposure limit for allergenic spores, but the differences in recommended limits vary as much as two orders of magnitude (e.g., ranging from 100 to 10,0000 CFU/m^3 for fungal spores). Part of the problem is the method involves viable spores only with an attempt to set standards for viable allergens where nonviable spores are contributors to the allergen symptoms. One report of workplace allergen concentrations may be of interest to some of the readers. See Table 3.8.

Table 3.8 Concentrations of Allergens in Some Workplace Environments

Workplace Environment	Thermophilic Actinomycetes	Total Fungi
Outside air	10	1,000
Domestic waste management	1,000	100,000*
Farming (normal activities)	—	10,000,000
Farming (handling moldy hay)	1,000,000,000	1,000,000,000**
Pig farms	1,000	10,000
Mushroom farms (composting)	10,000,000	100
Mushroom farms (picking)	100	100
Sugar beet processing	100	1,000
Cotton mills	100,000	1,000

* Mostly *Aspergillus* and *Penicillium*
** Mostly *Aspergillus*

Excerpted from Crook, B.[17]

An attempt is made herein to discuss some previously published suggested limits and comparative values. Their presentation is not to be taken as an endorsement by the author. These differences in interpretations are intended to assist in providing some additional insights.

1989 ACGIH Recommendations

Indoor exposures must be interpreted with respect to other environments (e.g., outdoor air and/or noncomplaint areas). Indoor mold spore levels should be similar to or lower than outdoor counts, and the genera of the viable spores should be similar in composition. Typically, the indoor fungal counts are filtered in the mechanical air handling units to half of the outdoor levels. If there has been an amplification source indoors either the total count of all the outdoor molds will increase, or the count of limited few will increase, altering the percentage composition of those few organisms in the total component.

They state that outdoor spore counts routinely exceed 1000 CFU/m^3 and may average near 10,000 CFU/m^3 during warmer, more humid months. With one exception, levels of any nonpathogenic or nonmycotoxin-producing fungi that are less than 100 CFU/m^3 have been of minimal concern. The exception is special environments, such as hospitals, where there are immune suppressed patients.

They reiterate that actinomycetes are typically in agricultural areas, not within indoor air environments. Thus, their presence is sufficient to indicate contamination.

Different Perspectives

There was a considerable change between the 1986 and 1989 interpretations offered by the ACGIH. The original 1986 limits are included herein, because some professionals continue to use these limits as guidelines. Then, too, there are those who choose different limits all together.

1986 ACGIH Recommendations

In 1986, the ACGIH stated that if the total count in an area of concern equals or exceeds 10,000 colony forming units per cubic meter, the levels are excessive and should be remediated. This number includes all sampled bioaerosols (e.g., mold spores, bacteria, and thermophilic actinomycetes).

They further stated that the presence of one fungus genera, exceeding 500 CFM/m^3, at the area of concern and not likewise found in the outdoor air sample(s), may indicate a need for remediation. The singular fungal colonies should be identified and enumerated. They included the following genera:

- *Acremonium*
- *Alternaria*
- *Aspergillus*
- *Aurobasidium*
- *Chaetomium*
- *Cladosporium*
- *Mucor*
- *Penicillium*
- *Stachybotrys*

The recommended limit for thermophilic actinomycetes is the same, 500 CFU/m^3. Identification and enumeration should be performed on the following bacteria:

- *Micropolyspora faeni*
- *Saccharomonospora spp.*
- *Thermoactinomyces spp.*
- *Thermomonospora spp.*

NIOSH and Experienced Allergy Consultants

The National Institute for Safety and Health (NIOSH) recommended in 1985 a limit of 1000 CFU/m^3 total of all viable microbes. They do not suggest a separate limit for mold spores and bacteria.

The American Academy of Allergy and Immunology has found that greater than 1000 CFU/m^3 total mold spores to be the mean threshold for allergy symptoms. Total refers to all viable and nonviable mold spores. If the viable count alone exceeds 1000 CFU/m^3 and the indoor comparisons appear in order, remediation may still be a consideration. These are primarily for nonpathogenic, allergenic fungi and bacteria.

Some environmental professionals choose to use 200 CFU/m^3 viable microorganisms as the minimum indicator for probable indoor fungal amplification and bacterial enhancement. Increased numbers of commensal bacteria may indicate high occupant density and/or poor building ventilation.

One study suggested that spore types should be a major consideration. Symptoms had been reported at levels of 100 spores/m^3 (genus *Alternaria*) and 3000 spores/m^3 (genus *Cladosporium*).[18] This study was, however, performed for total spores, not just viables, which could suggest lower limits for viable genera.

Some professionals have noted that exposures in excess of 1000 CFU/m^3 total microbes have been found in noncomplaint areas, and others suggest that inconsistencies between complaint areas and other noncomplaint/outdoor areas (particularly where the microbes have been identified as known allergens) are more reliable.

In one study involving a comparison of complaint and noncomplaint areas, levels as low as 65 CFU/m^3 of *Stachybotrys* and *Aspergillus versicolor* were identified only in the complaint areas. The outdoor air had different genera. Consequently, remediation was recommended.[19] The issue here was not the total count but the amplification of specific genera over that which was measured outdoors, not total counts. Yet, a count resulting in 65 CFU/m^3 would be only two colonies on a nutrient agar plate as sampled by an Andersen impactor for one minute. The collectate may be ruled out as an anomaly due to the limited number of samples taken. Thus, when costly recommendations are possible and low counts likely, it is advisable to take a few more samples at any one given location or perform confirmation sampling at a later time.

The limits posed herein are an attempt to provide the reader with some line of logic or point at which careful consideration of other factors should be examined. As with statistics, the greater the number of samples, the more reliable the results. While comparison sampling is a sound scientific approach for ascertaining amplification, viable sampling provides only a piece of the picture. An easily referenced standard for acceptable mold and bacteria spore levels is not "just over the horizon."

REFERENCES

1 Smith, E. Grant. *Sampling and Identifying Allergenic Pollens and Molds: An Illustrated Identification Manual for Air Samplers.* Blewstone Press, San Antonio, Texas, 1990. p. 43.

2 ACGIH. Bioaerosols: Airborne Viable Microorganisms in Office Environments—Sampling Protocol and Analytical Procedures. *Applied Industrial Hygiene*, April 1986. p. R-22.

3 Cole, Garry T. and Harvey C. Hock. *The Fungal Spore and Disease Initiation in Plants and Animals.* Plenum Press, New York, New York, 1991, p. 383.

4 Al-Doory, Yousef and Joanne F. Domson. *Mould Allergy.* Lea & Febiger, Philadelphia, Pennsylvania, 1984. pp. 36-37.

5 Ibid. pp. 287-288.

6 Crissey, John T., MD, Heidi Lang, and Lawrence Parish. *Manual of Medical Mycology.* Blackwell Science, 1995, pp. 190-201.

7 Smith, E. Grant. *Sampling and Identifying Allergenic Pollens and Molds: An Illustrated Identification Manual for Air Samplers.* Blewstone Press, San Antonio, Texas, 1990. p. 16.

8 Ibid. p. 42.

9 Al-Doory, Yousef and Joanne F. Domson. *Mould Allergy.* Lea & Febiger, Philadelphia, Pennsylvania, 1984. pp. 24-25.

10 Ibid. pp. 235-237.

11 ACGIH. Bioaerosols: Airborne Viable Microorganisms in Office Environments--Sampling Protocol and Analytical Procedures. *Applied Industrial Hygiene*, April 1986. p. R-22.

12 ACGIH. *Guidelines for the Assessment of Bioaerosols in the Indoor Environment.* American Conference of Governmental Industrial Hygienists, Cincinnati, Ohio, 1989.

13 Ibid. p. R-19.

14 ACGIH. *Guidelines for the Assessment of Bioaerosols in the Indoor Environment—Air Sampling.* American Conference of Governmental Industrial Hygienists, Cincinnati, Ohio, 1989. p. 1.

17 Crook, B and J. Lacey. Airborne allergenic microorganisms associated with mushroom cultivation. Grana. 30:445 (1991).
18 Dhillon, M.. Current status of mold immunotherapy. *Ann. Allergy.* 66:385 (1991).
19 Morey, Philip R., Ph.D. Controlling Microbial Contamination to Prevent Building-Related Illness and Remediation Costs. [Newsletter] Clayton Environmental Consultants, Edison, New Jersey, July 1993,. 15(2):3.

Chapter 4

AIRBORNE PATHOGENS/ MICROBIAL TOXINS

Pathogenic microbes must be viable in order to cause disease where the toxins may be associated with viable and/or nonviable organisms. The pathogens cause disease by attacking specific tissues, multiplying and invading. Microbial toxins are metabolic by-products of fungi or components of the outer membrane of bacteria. In each instance, there is a dose-response relationship similar to toxic chemical exposures. The dose is oftentimes undetermined, and the response depends on individual susceptibility and pathogenicity of the organism.

Although there have been some efforts to sample for the presence of pathogens and microbial toxins, airborne sampling is complicated, controversial, and expensive. Most efforts have involved bulk and surface wipe sampling.

Sampling further requires lengthy culture times, complicated by the presence of other environmental, nonpathogenic microbes which may mask and prevent identification of the pathogen by over-growing the culture medium. Then, when a suspect colony is isolated, additional time is required to perform more extensive, complicated analyses/cultures in order to confirm its identity and, in many cases, to determine the species and strain.

Whereas many of the pathogens are potentially lethal, the time required for identification is frequently not practical. For this reason, the presence of a pathogen is often assumed where patients are symptomatic, reacted upon later. Typically, health professionals attempt to isolate the source environment through studying sick patients' habits and environmental associations, trying to recreate a common denominator for an observed epidemic.

Although pathogenic diseases are frequently occupation and area depend, the greatest attention to pathogen spread occurs in hospitals. An infected patient aerosolizes (e.g., coughs or sneezes) viable pathogens, and immune suppressed patients are particularly susceptible aerosolized pathogens and nonpathogens. Immune suppressed patients include those with AIDS, organ transplants, chemotherapy treatments, and diabetes. Operating rooms and invasive diagnostic testing areas are possibly involved. Infants and the elderly can be compromised.

Those weakened by living conditions (e.g., the homeless) and inadequate nourishment (e.g., Somalians) have an enhanced susceptibility to pathogens.

Then, too, some pathogenic microorganisms are capable of causing disease in healthy individuals under special circumstances. Contaminated intravenous solutions and hospital implements provide an easy avenue for entry. Puncture wounds in contaminated environments are fertile soil, and damaged tissues propagate invasive challenges. Opportunity for pathogenic assaults is subsequently dependent upon the environment.

Some pathogens multiply in air handlers and water reservoirs. Many are contained within environmental dust. The problems are rarely identified prior to the spread of disease to others associated with the dispersing mechanism. Sampling is generally requested after a problem has arisen. Proactive sampling in environments where specific pathogens may thrive is indicated.

As invasion and amplification require viable pathogens, this chapter addresses methods for identifying the "viable" microbes, requiring lengthy tests. Other methods, discussed in Chapter 6, are available which can identify the pathogens rapidly, accurately. However, the latter methods also identify nonviable/noninvasive microbes, and results are sometimes difficult to interpret. The method of choice may be a combination of approaches, depending upon the situation. Professional judgment and collaboration are paramount. Understanding the context of these two chapters should provide considerable assistance.

AIRBORNE PATHOGENIC FUNGI[1–3]

Pathogenic fungi include those which cause disease through inhalation or airborne exposures and those which are more opportunistic in initiating disease through puncture wounds, open sores, cuts and abrasions, and moist dermal conditions, an extensive list of species. They range from localized, surface discomfort to whole body invasion and death.

The primary focus of this section is to discuss the airborne fungi which can invade and cause death. Only the principal pathogenic molds are discussed. Those which are infrequently or not typically found in the United States may require additional research, strategy, and sampling development. The following information should provide direction and guidance to addressing potentially lethal environmental situations.

Disease and Occurrence of Major Pathogenic Fungi

Understanding the diseases caused by specific pathogenic fungi and regions/areas where they primarily occur is necessary to determine a need for sampling. At the same time, however, the environmental professional must not be limited by this information. An outbreak may be atypical.

Various Species of Aspergillus

Several species of *Aspergillus*, primarily the *Aspergillus fumigatus* and *flavis,* can cause a disease commonly referred to as "aspergilliosis." As the genus *Aspergillus* is one of the more common fungal spores found in air monitoring, the pathogenicity is minimal, unless an individual is immune suppressed or has a debilitating illness. The initial symptoms may be localized to the lungs, ears, or perinasal sinuses. Once the fungus has found a place to grow, the hyphae may grow into the blood stream to deposit spores to be disseminated to other parts of the body. The result may be the formation of abscesses or granulomas in the brain, heart, kidney, and spleen.

As the fungus thrives on those who are immune suppressed, the greatest areas of concern are in the hospitals, clinics, nursing homes, and hospices. Immune suppressed patients include AIDS patients, organ transplant patients (e.g., heart transplants), premature infants, and radiologically-treated leukemia patients. Invasive diagnostic suites and operating rooms are particularly critical. The presence of even low levels of the invasive forms of certain *Aspergillus* species can be deadly. The hospitals seek to control these environments. See Table 4.1 for historic outbreaks.

Some studies have included several fungal species to the list of potential invaders. One such study includes the following:[4]

- *Aspergillus fumigatus*
- *Aspergillus flavis* (also producers the carcinogen aflatoxin)
- *Aspergillus niger*
- *Aspergillus terreus*
- *Petriellidium boydit*
- *Fusarium spp.*
- *Mucoraceae spp.*
- *Phoma spp.*
- *Alternaria spp.*
- *Penicillium spp.*
- Others

Other studies include *Zygomycetes* and *Rhizopus*.[5,6] Most agree, despite the failed concurrence of opinions, that *Aspergillus fumigatus* and *flavis* are the predominant fungal spores of concern in hospitals. The other species should be considered "potentially invasive" by the environmental professional.

There are a few occupational exposure concerns to the genus *Aspergillus* as well. Aspergilliosis was first described by an Italian in the 1700s as an occupational disease of people who handled grains and birds. This concern, however, has lost its significance in modern times. As the fungus grows best on compost and damp organic matter, some feel that farmers and field workers are at some risk.

Table 4.1 Historic Outbreaks of Nosocomial Aspergillosis

Number of Patients	Type of Disease	Host	Environmental Factors
7	not stated	bone marrow transplants	recent construction and defective air conditioners
32	pneumonia/sinusitis and colonization	respiratory patients	building construction and defective ventilation units
10	invasive pulmonary disease	immunosuppressed (7) malignancy (2) elderly (1)	road construction and defective air conditioners
9	invasive pulmonary disease	not stated	building construction
3	invasive pulmonary disease (2) and colonization (1)	renal transplants	hospital renovation
8	invasive pulmonary with one dissemination	acute hematological malignancy, neutropenia	wet fireproofing and dust on false ceiling
4	invasive pulmonary with one dissemination	renal transplantation immunosuppressive rx	pigeon excreta external through the air intake

Excerpted from Nosocomial Aspergillosis: An Increasing Problem.[7]

Histoplasma capsulatum

Histoplasmosis, caused by inhalation of the mold spores of *Histoplasmoa capsulatum*, results in a variety of clinical manifestations. Primary histoplasmosis is a common, region oriented, benign disease of the lungs involving a mild or asymptomatic pulmonary infection with a cough, fever, and malaise. Chronic cases may go on to experience a productive cough, low-grade fever, and a chest X-ray showing cavitation. As the cavitation resembles tuberculosis, however, many cases are misdiagnosed. Then, a small percent (less than 1 percent of all cases) of those showing symptoms develop in the progressive form which can be lethal.

Progressive, systemic histoplasmosis is less common. It is characterized by emaciation, leukopenia, secondary anemia, and irregular pyrexia. There is frequently ulceration of the naso-oral-pharyngeal cavities and intestines with generalized infection of the lymph nodes, spleen, and liver.

There have been reports of *Histoplasma capsulatum* related chronic meningitis in non immune suppressed hosts, and some have speculated that the fungus also contributes to persistent or recurrent carpal tunnel syndrome.[8,9] Symptoms in immune suppressed patients are highly variable, and the fungus is generally associated with other pathogens.

Within the United States, *Histoplasma capsulatum* is endemic in the Central Mississippi Valley, Ohio Valley, and along the Appalachian Mountains. See Figure 4.1. Other endemic areas include sections of South America and Central America, and there are occasional outbreaks in various parts of the country and world. It is thought that the spores become airborne, and individuals become infected by inhaling *Histoplasma capsulatum* from the air. Disease severity in a healthy adult is generally dose-related.

The growth of the fungus appears to be associated with decaying or composted manure of chickens, birds (especially starlings), and bats (e.g., "cave disease"). Thus, the potential for elevated exposure potential occurs predominately around the following areas:

- Soil in and around chicken houses or silos that have not been disturbed for a long time.
- Soil under trees that have served as roosting places for starlings, grackles, or other birds.
- Farm land where fertilizers or soil-containing chicken droppings or large quantities of organic matter have been spread.
- Dust around a building being demolished which is where pigeons had previously roosted.

Contaminated soils and dust are the greatest sources of exposure to the drought resistant spores. Due to the analytical costs, proactive sampling is rarely performed. In one recent laboratory quote, the minimum analytical cost was

$4000. The reason for the high cost is that analysis involves animal studies with at least ten samples.

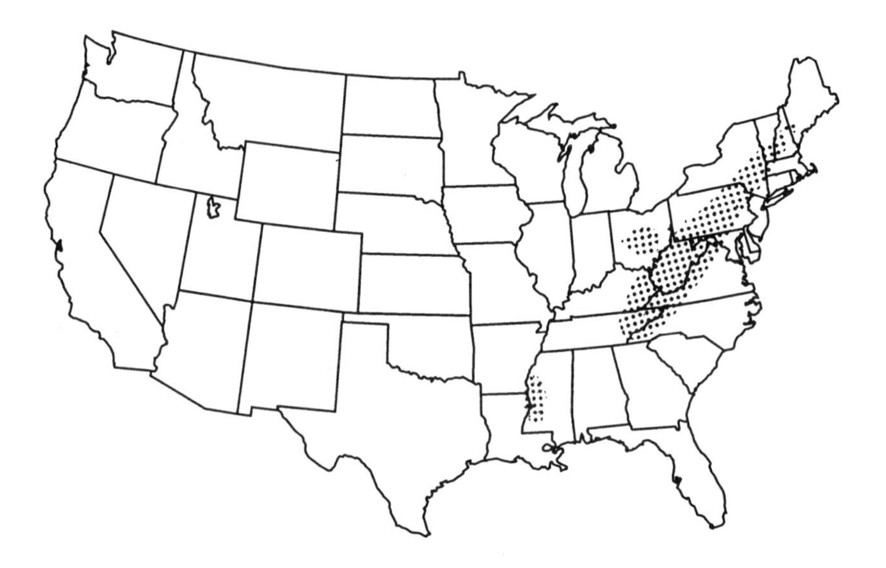

Figure 4.1 Map of United States depicting the states/areas where the occurrence of "Histoplasmosis" is known.

Sampling is, however, sometimes performed where disease is already known. The extent and impact of the associated disease are increasing, and corrective actions are initiated. Still, sampling is generally performed only after clean-up or upon completion of corrective actions. Samples are taken for confirmation of adequacy in the response program.

Coccidioides immitis

Coccidioidomycosis (often referred to as "Valley fever"), a disease caused by inhalation of the mold spore of *Coccidioides immitis*, is generally an acute, benign, and self-limiting respiratory infection. It is characterized by chills, fever, cough, and localized chest pain. Less frequently, the disease involves the visceral organs (e.g., bone, skin, and subcutaneous tissue) where abscess may develop. Illness usually lasts for two to three weeks, leaving some scarring of the lungs. However, if an individual has had a previous sensitizing exposure, the symptoms can be lethal.

In the more advanced stages of the disease, abscess formation in the lungs and the rest of the body, including the central nervous system, may occur. Immunologic resistance generally develops in those who recover.

Organ involvement in immune suppressed patients includes the skin, bones, joints, and central nervous system. It may or may not be associated with other pathogens.

The disease is endemic in hot, arid/semiarid areas of the southwestern parts of the United States, northern Mexico, and South America. Endemic regions in the United States include sections of Arizona, California (e.g., San Joaquin Valley), Nevada, New Mexico, Texas, and Utah.[10] See Figure 4.2. The fungal spores become airborne, and disease is thought to be dose-related.

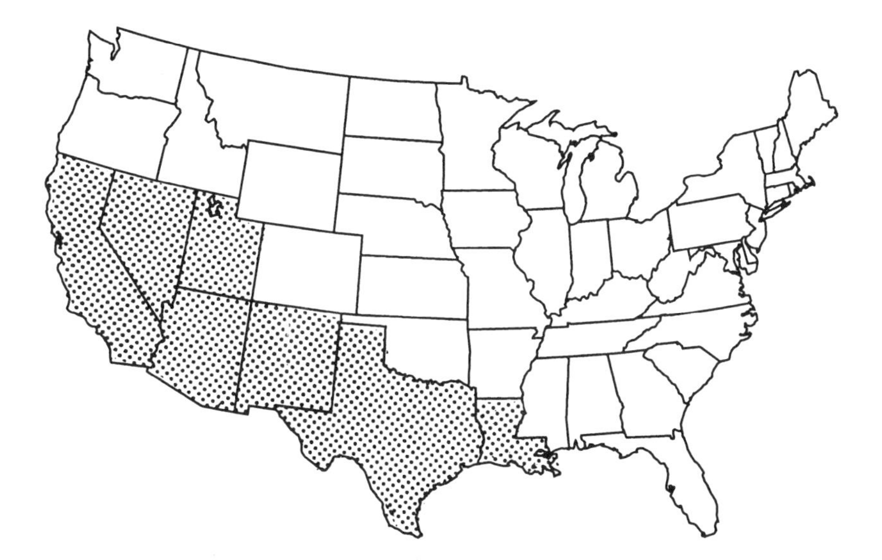

Figure 4.2 Map of United States depicting the states/areas where the occurrence of "Coccidioidomycosis" is known .

Contaminated dust is the principal source of exposure. So, workers and households in arid or semiarid regions have a higher risk of contracting the disease than others. Most of the infections occur during the dry season, following dust storms. Occupations of greatest risk include migrant workers, farmers, construction workers, military personnel, and road workers. Clinical laboratory exposures may also occur due to inadequate controls of the highly infectious, amplified cultures.

Sampling should be in dusty occupational environments, in regions endemic with *Coccidioides immitis.* With a few minor exceptions, the sampling methodology for this fungus is similar to that of *Histoplasma capsulatum.*

Cryptococcus neoformans

Cryptococcus neoformans attacks the central nervous system, producing a subacute or chronic meningitis. It may also impact the lung tissue and cause low grade inflammatory lesions which may be mistaken for tuberculosis or a neoplasm. Occasional cases arise whereby the fungus gives rise to localized nodular lesions of the skin or generalized infections with lesions of the skin, bones, and viscera.

Rare in healthy adults, *Cryptococcus neoformans* varieties *neoformans* and *gattii* are reported to be the cause of death in 6 to 10 percent of the HIV patients. Thus, they are the most common AIDS-related, lethal mycoses.[11] These strains have been known to cause blindness in AIDS patients and are the foremost cause of central nervous system infections in the same. Diabetics, cancer patients, and recipients of organ transplant patients are at risk of generalized infection. Fungal infections range from 5 percent in recipients of kidney transplant patients to as high as 40 percent among recipients of liver transplants.[12]

The most commonly reported source of contaminated dust is pigeon droppings. There are also numerous reports from Australia that the source of infections can be attributed to certain species of Eucalyptus (e.g., River red gum and Forest red gum).[13]

Prior to an incident or outbreak, routine sampling for fungi, identifying *Cryptococcus neoformans,* is indicated especially in areas where immune suppressed patients may be compromised the greatest (e.g., operating rooms). Particular attention should be given to the air handler systems (e.g., growths occurring beyond the filtration devices) supplying the operating rooms, minor surgery suites, and invasive diagnostic rooms.

Cryptococcus neoformans occurs in cultures in the form of round yeast-like cells, surrounded by a large gelatinous capsule which is though to contribute to its ability to resist phagocytosis, bypass the initial body defenses. Although they measure 5 to 20 microns in diameter when in tissues, the cells measure 2 to 5 microns in culture media.

Other Pathogenic Fungi

Most of the other fungi which have not been mentioned are either nonpathogenic, rare occurrences of disease in the United States, or rarely infectious through inhalation. Some of the nonairborne infectious fungi are listed in Table 4.2.

It should also be noted that health and comfort concerns may be associated with volatile organic by-products of some of the fungi. Some of the published volatile organic compounds are listed in Table 4.3. Bacteria and fungi produce some of the same volatile organics. For example, the most common ones are acetone, 2-butanol, dimethyl trisulfide, methanol, and 1-propanol. Levels and their contribution to the total organic exposures indoors have yet to be determined.[14]

Table 4.2 Nonairborne Infectious Fungi[15]

Genera	Disease	Route(s) of Entry
DERMATOPHYTOSES (involve the sking, hair, and nails)		
Microsporum	tinea capitis	combs, brushes, and barber's clippers
Trichophyton	ringworms	surface transference
Epidermophyton	epidermophytosis	surface transference
SUPERFICIAL MYCOSES (involve the skin and subcutaneous tissues)		
Sporotrichosis	sporotrichosis	puncture wounds
Cladosporium carrionii	chromoblastomycosis	puncture wounds
Phialophora verrucosa	chromoblastomycosis	puncture wounds
Fonsecaea pedrosoi	chromoblastomycosis	puncture wounds
Nocardia	maduromycosis	skin injuries
Streptomyces	maduromycosis	skin injuries
Allescheria boydii	maduromycosis	skin injuries
OPPORTUNISTIC INVADOR (involves mouth, vagina, skin, nails, bronchi, or lungs)		
Candida	candidiasus	surface transference

The genera *Penicillium* and *Ulociadium* produce organic by-products which have been reported to cause headaches and eye, nose, and throat irritation, respectively.[16] The organic by-products, or volatile organic components, require further research, and there are no published practical methodologies for air sampling and interpretation in their presence.

Table 4.3 Volatile Organic Compounds Associated with Fungi and Bacteria[16]

MOST FREQUENTLY REPORTED	
3-Methyl-1-butanol	Heptanone
	Indole
1-Octen-3-ol	Hexanal
	Hexanol
3-Octanol (1-octan-3-ol)	Methanol
3-Octanone	4-Methoxybenzaldehyde
	Nonanone
Acetone	2-Octen-1-ol

Table 4.3 (continued)

Dimethyl disulfide	Phenol
Dimethyl trisulfide	Trimethyl amine
Ethanol	Benzaldehyde

Sampling and Analytical Methodologies for Fungi[17,18]

Although many are drought resistance, most pathogenic fungi do not retain their viability during air sampling or may be over grown on the nutrient media by other fungi, masking the presence of the slower growing pathogens. Where air sampling has been performed for a total spectrum of fungi, however, a pathogen may inadvertently become identified. In accordance with the ACGIH, the air sampling equipment most likely to result in such identification are the sieve impactor (e.g., Andersen single-stage impactor) and a slit impactor (e.g., Burkhard sampler) with collection efficiencies to one micron in diameter.

Swipe or bulk sampling may be performed of suspect dust, surface areas, and source material (e.g., medical implements). These methods can be used for all forms of pathogenic microbes/toxins and are, therefore, described at the end of this chapter under the heading of Diagnostic Methodologies.

As with allergenic molds, either a malt extract or Rose Bengal extract can be used as the nutrient growth media. With a few exceptions, most can be grown in similar room temperature conditions and light requirements, and the culture plates should checked daily, read after five days. Exceptions include the following known pathogenic fungi:

- *Aspergillus fumigatus* is thermotolerant and will grow at temperatures of 45° C, or higher, which is an extreme temperature for most fungi. The temperature serves as a means for limiting other growths.
- *Cryptococcus neoformans* grows best at 37° C.

Once a pathogenic genus has been identified, species and sometimes variety (e.g., *Cryptococcus neoformans* variety *neoformans* and variety *gattii*) determinations require additional plating (e.g., slant specialized nutrient agar in test tubes). This requires a trained laboratory technician and may involve several steps with additional culturing, plating, replating, and stains. These additional efforts may be time consuming and expensive.

Interpretation of Results

There are no definitive guidelines for interpretation of results. The ACGIH suggests that airborne levels of less than 100 colony forming units per cubic meter (CFU/m^3) are not a concern, except in environments where there are immune suppressed people. These limits may not be adequate for the pathogenic types.

Where a pathogenic fungus is region dependent, baseline levels in symptom-free areas of endemic regions should provide a means for dose comparisons. There are presently no widely published findings nor recognized research studies which have identified baseline levels and suggested response limits.

For immune suppressed patients, any level of exposure may be potentially lethal. Thus, its confirmed presence in a hospital treatment area or association with medical implements should constitute reason for concern. The source should be found and remedied, medical implements cleaned.

With immune suppressed patients, there has been no determination as to an acceptable level. Many health/environmental professionals express concern at any level of identification and remediate to zero detection.

Limitations of Sampling and Analytical Methodologies

The lack of established standards limits one's ability to interpret the results, and other problems are associated with the need to respond rapidly. Culturing takes up to five days or longer, time which may be valuable wherein a problem has already arisen. The time factor may be particularly important in hospitals. Immune suppressed patients may be exposed, such as during invasive check-ups or operating rooms. Time may compromise patients or close down a vital area of a hospital until the results have become available and recommendations completed.

MYCOTOXINS[14,19]

Mycotoxins are toxic by-products of the metabolic process of certain fungi. Fungal toxins are produced by any of a number of fungi, ranging from edible poisonous/hallucinogenic mushrooms (or mushroom toxins/poisons) to inhaled or consumed mold components which have the mycotoxin. Some of these components include, but are not limited to, mycelium, spores, and molded substrates/food sources. The dose-response is based upon the type of toxin, the animal species impacted, and the route of entry.[20]

At the present, the greatest area of concern for mycotoxins is in the farming business due to the impact of many of these toxins on the ability of livestock to convert food, fight off disease, and reproduce. Minimal attention has been brought to bear on airborne mycotoxin exposures to man.

Disease and Occurrence

Aspergillus flavis and *Aspergillus parasiticus* create aflatoxins which are carcinogenic and may cause acute health effects. These molds grow predominately in warm, humid environments, and are typically associated with peanuts, grains, sweet potatoes, corn, peas, and rice. Aflatoxins have been reported to cause liver carcinomas in animals and are believed to be amongst the most potent of carcinogens.[21] Although some studies have concluded that inhaled aflatoxin does not appear to be associated with lung cancer, there is one report of two deaths due to pulmonary adenomatosis. In the latter report, autopsies indicated the presence of aflatoxin in their lungs. The findings, however, were not conclusive.[22]

Trichothecene toxins are a by-product of the metabolism of *Fusarium*, *Acremonium*, *Trichoderma*, *Myrothescium*, and *Stachybotrys*. Reported symptoms include headaches, sore throats, hair loss, flu-like illness, diarrhea, fatigue, dermatitis, and generalized malaise. In one instance, occupants in a household in Illinois were found to have symptoms suggestive of "trichothecene toxicosis." The cause was identified by air sample collection, using an electrostatic precipitator that collected toxin from the air handling system which was heavily infested with *Stachybotrys atra*. The mold was later found in the air ducts and damp fiberboard. A thorough cleaning rectified the health complaints of skin and respiratory system irritation.[23]

Some species of *Penicillium* produce several forms of toxins. One such species is *Penicillium viridicatum*. Not all species of *Penicillum* apply. Thus, where the genera *Aspergillus* and *Penicillium* are identified, the species should also be determined. Species determination may be required in other cases as well.

To complicate matters, toxins are not always produced. The nutrient (or substrate) upon which the mold grows impacts its metabolites. Indoor environments and laboratory nutrients may result in a variation from the norm. Thus, the presence of amplified numbers of a mycotoxin producing mold in an air handler does not necessarily follow that it is producing the toxin. Likewise, the failure of a mold to produce toxins in laboratory conditions does not necessarily mean that it does not produce the toxin in outdoor environments.

Most cases of mycotoxin poisoning appear in rural and agricultural settings as a result of ingestion and/or skin contact. There have been a few reports of cancer related to consumption of aflatoxins in humans. Yet, all-and-all, there have been minimal studies and research into the possible relationship between airborne aflatoxins and its impact on humans.

None of the molds known to produce mycotoxins are typical in the outdoor air environments. They may, however, become amplified indoors where conditions may be more conducive to growth. These mycotoxins readily grow in environmental conditions similar to that of allergenic molds.

Sampling and Analytical Methodologies for Mycotoxins[19]

There are two different approaches to sampling for mycotoxins. The fungi may be identified with emphasis on those which are know to create toxic byproducts and total spores counted, or the toxins themselves can be sampled.

Where the fungi are to be identified from culturing, many of the toxin-producing fungi may not compete well with the other fungi, or they may no longer be viable where the toxins still persist in nonviable fungi. For this reason, a total count of all viable and nonviable spores must be performed, along with culturing for identification purposes.

Toxin-Producing Fungi Identification

Culture impactors (e.g., Andersen single-stage impactor and Burkhard sampler) are the equipment of choice for viable fungal samplers. The nutrient media of choice is either a malt extract or Rose Bengel agar. They can typically be grown at room temperature with similar light requirements to that of the allergenic fungi. Plates should be checked daily and read after five days of incubation. Once a genus of concern has been ascertained, species determination and/or verification will require additional laboratory analyses (e.g., additional incubation and special tests).

Total counts should be made using a separate sampler, a particulate impactor (e.g., Burkhard sampler with treated microscope slides). This method is similar to that of total allergenic spore counting procedures. In some cases, spores can be identified by genus (e.g., *Stachybotrys* are large grenade-shaped in appearance). Some counts may be by genus. Also, the collection efficiency (as compared to the impactors) must be known and a correction factor applied to the results.

Toxin Identification

The methodologies for sampling airborne mycotoxins are available for a few of the better known toxins (e.g., aflatoxins and some trichothecenes), but commercial laboratories typically do not perform the required analyses due to low volume requirements and the cost of developing a new, seldom required method.

In response to the U.S. Food and Drug Administration (FDA) requirements that aflatoxin levels in food be less than 20 parts of aflatoxin per million parts of food substance (ppm), some owners of granaries and corn storage facilities routinely perform a quick colorimetric test for the presence of aflatoxin. Although the FDA probes for larger samples and has analyses performed by a laboratory, the colorimetric test kits (based on immunoassay tagging of the mycotoxin) are available commercially and are readily used for screening by the storage facilities.[24] This approach has yet to be introduced into and/or used by environmental professionals to screen air samples.

Interpretation of Results

There are no definitive guidelines for interpretation of results. The environmental professional is entirely on his own or must rely on guidance from one with experience in this area of expertise. The laboratory chosen to perform the work should be able to either provide guidance or direct you to someone competent in that area. Each case will be unique and depend on several variables including intent and sampling strategy. All factor must be considered in the final interpretation.

Limitations of Sampling and Analytical Methodologies

Sampling and analytical limitations include difficulty of fungal species identification; the use of two sampling techniques and need to interpret them with no guidelines; and the lengthy analytical time required for identification may further complicate an already compromised situation. Most of the sampling done at the present is in the research arena, and laboratories prepared to provide the analytical services and expertise may be difficult, if not impossible, to find.

AIRBORNE PATHOGENIC BACTERIA[25–27]

Most bacteria are ubiquitous, nonpathogenic. Yet, airborne pathogens are a concern in certain occupations and in indoor air environments. Most of these pathogens are aerosolized from soil, water, plants, animals, and people. Outdoor exposures may also occur and are generally associated with windblown dust. Indoor bacteria are generally human commensal, and the airborne pathogens usually infect the respiratory tract. Enormous numbers of moisture droplets are expelled during coughing, and smaller amounts are expelled through talking. A single sneeze may generate as many as 10,000 to 100,000 bacteria. On the more positive side, most bacteria do not survive once they have become airborne.

Each of the pathogenic bacteria has its own distinct survival adaptation which permits it to survive long enough to invade its target host. The bacteria *Legionella* is highly resistant to acids. The thicker walled, Gram-positive bacteria (e.g., diphtheria) are resistant to drying, and one bacteria forms protective spores (e.g., genera *Bacillus*). The more prominent airborne pathogenic bacteria found within the United States, those posing the greatest concerns in the occupational and indoor air environments, are discussed herein.

Pathogenic *Legionella* [28,29]

Diseases caused by the bacteria genus *Legionella* are estimated by the Centers for Disease Control at between 25,000 and 100,000 per year in the United States alone. Collectively, these diseases are referred to as "legionellosis" (e.g., Legionnaires' disease and Pontiac fever).

There are thirty-four known species of *Legionella*, fifty known serogroups. Although many of the species have not been implicated in human disease, *Legionella pneumophila*, Serogroup 1, is the most deadly, most frequently implicated form associated with Legionnaires' disease. The latter Serogroup 1 and many of the others are frequently found in all environments and water reservoirs. Their presence alone is not sufficient to constitute a threat.

Virulence is related to the following:

- Species and serogroup
- Total amount of viable bacteria in a water reservoir
- Aerosolization of contaminated water
- Distribution of aerosolized droplets to human hosts
- Cooling towers and evaporative condensers—least potential
- Water heaters and holding tanks
- Pipes containing stagnant water
- Faucet aerators
- Shower heads
- Whirlpool baths
- Humidifiers and foggers—highest potential
- Reduced host defenses

Other sources of *Legionella* have been identified in unique environments. In one case, an outbreak was tracked back to the misting of produce in a Louisiana grocery store.[30] Another involved a five patient outbreak which was linked to the aerosol from a decorative fountain associated with a private water supply in an Orlando, Florida hotel.[31]

The warmer the water reservoir, the greater the potential for amplification of the viable organisms. Viable organisms have also been isolated from stagnant pools, lakes, and puddles of water.[32] An example of a highly virulent situation is the presence of *Legionella pneumophila* in large amount, aerosolized and distributed from a warm humidifier into the living environment of an immune suppressed, elderly patient.

Although Legionnaires' disease is primarily a disease of the elderly, it may impact younger patients whose health has been compromised and immune suppressed patients (e.g., AIDS patients). In a healthy adult of working age, infections are usually asymptomatic and may in some cases result in mild symptoms of headache and fever. In the elderly, prior to the onset of pneumonia, intestinal disorders are common, followed by high fever, chills, and muscle aches. These symptoms progress to a dry cough and chest/abdominal pains. Where death occurs, it is typically due to respiratory failure. Death occurs to only fifteen percent of the less than five percent of the exposed individuals showing disease which develops within three to nine days of exposure.

Pontiac fever is not quite so pathogenic. Within two to three days of exposure, approximately 95 percent of all exposed individuals develop short-term flu-like symptoms.

The *Legionella* bacteria is a thin, Gram-negative rod with complex nutritional requirements. A unique characteristic which is used in isolating the genus is its ability to survive at pH 2.

Sampling and Analytical Methodologies for Legionella

Legionella pneumophila is fragile. Once aerosolized and dehydrated, it loses its viability. For this reason, air sampling is unreliable, not recommended. Microbe damage, and loss of viability, may result in diminished counts.

Water sample collection procedures have been developed. They are considered more reliable and a means for interpreting the results has been developed. Whenever possible, one liter of water should be collected in a sterile container. Screw-capped, plastic bottles are preferred collection containers. If the water source has recently been treated with chlorine, neutralize the sample(s) with 0.5 milliliters of 0.1 N sodium thiosulfate. Several samples should be taken from suspect/potential reservoirs.[33]

Sometimes, a liter of water cannot be collected. Collect as much as is feasible, and place it in an appropriately sized container. A one milliliter sample may evaporate from or become irretrievable from a one liter container. Where sampling from a faucet or shower head reservoir, available water is minimal. In such instances, the reservoir should be swabbed with sterile swabs (e.g., polyester medical swabs with a wood stick), contained within a sterile enclosure. Upon sampling, each sample swab should be submerged in a small amount of water from the source outlet (e.g., shower, faucet, etc.).

Upon completion of sample collection, all samples should then be sent by overnight freight to a laboratory for analysis. If this is not possible, the samples should be refrigerated until they can be processed. Otherwise, avoid temperature extremes both in storage and shipping.

At an experienced laboratory, the known aliquots of collected water sample will be plated onto nutrient medium and cultured for up to three weeks. The nutrient medium of choice is enhanced, buffered-charcoal yeast extract agar. Prior the plating, many of the sample are acid-treated for 15 to 30 minutes, then neutralized. Most of the other non*Legionella* microbes are destroyed, and the *Legionella*, having retained its viability, grows unrestricted.

The plates are incubated at 35° C and examined daily for ten days. *Legionella* bacteria are slow growing organisms, and colonies that develop early are not likely to be *Legionella* bacteria. Sometimes, where there are excessive growths, a determination may be made within a few days. Otherwise, the growth may not appear for three weeks. Results are reported in terms of organisms per milliliter of water.

Interpretation of Results

Exposure action levels for contaminated water have been developed by researchers. These levels are based upon extensive experience and field studies. They determined the levels of organisms present in various water sources where Legionnaires' disease was known and where it was not a reported problem. The data was compiled and numbers compared. It is from these studies that the action levels were created in order to anticipate and avoid costly outbreaks. These suggested limits are provided within Table 4.4. They are intended to be used as "guidelines only."

Table 4.4 Suggested Remediation Action Criteria Levels for *Legionella*

Legionella		ACTION LEVELS	
Organisms per Milliliter	CT/EC	Potable Water	Humidifier/Fogger
< 1	1	2	3
to 9 2	3	4	
to 99	3	4	5
to 999	4	5	5
> 1,000	5	5	5

CT/EC: Cooling towers and evaporative condensers
Excerpted from "Reducing Risks Associated with *Legionella* Bacteria in Building Water Systems."[26]

The action levels are 1 through 5. Action Level 5 requires the greatest amount of care (e.g., immediate cleaning and/or biocide treatment). Each suggested action includes the preceding actions as well. The minimum recommendation for Action Level 1 is the review of routine maintenance programs as recommended by equipment manufacturers. Action Level 2 is a follow-up review for evidence of *Legionella* amplification. Action Level 3 represents low contamination, yet, elevated levels of concern. A review of the premises for direct and indirect bioaerosol contact with occupants and their health risk status. Where outbreaks may become possible, Action Level 4 suggests cleaning and/or biocide treatment of the equipment. See Table 4.5 for specific remedial actions.

Limitations of Sampling and Analytical Methodologies

Legionella may be hidden and amplified within the cells of other microorganisms (e.g., protozoa). Thus, a negative result does not necessarily indicate the environmental source of a sample is free of *Legionella*. In such cases, the environmental professional should comment that low levels are an indication of "low risk." There are no absolutes.

The occurrence of a potential misdiagnosis of disguised, elevated levels has led to a choice not to sample by some environmental professionals. However, this choice may potentially result in not identifying those which are elevated.

In a 1991 incident involving a Richmond, California, Social Security Office Building, two people died of Legionnaires' disease. The outbreak involved ten cases and created considerable negative publicity. On March 13, 1995, the U.S. Department of Justice settled the case rather than go to trial. The amount of the settlement was not disclosed, though it was apparently substantial. A key question in the case was, "Does the policy of not testing the waiting for cases of Legionnaires' disease to occur, provide a reasonable standard of care?"[34]

Another incident involving exposures in a Jacuzzi on a cruise ship occurred in 1994. A more recent event involved misting of grocery produce and a link to Legionnaires' disease. Each of these incidents may have been avoided had prior sampling been performed, elevated levels identified, and the sources remediated.

The criteria levels described above were developed by a laboratory that maintained a stringent quality assurance program, including in-house proficiency testing of the laboratory personnel to insure accuracy and reproducibility. The interpretative value of these data may not be applicable with quantitative values from other laboratories that do not have similar quality standards.

Table 4.5 Suggested Remedial Response Action

Action Level	Suggested Remedial Response
1	Review routine maintenanceprogram recommended by the manufacturer of the equipment to ensure that the recommended program is being followed. The presence of barely detectable numbers of *Legionella* represents a low level of concern.
2	Implement Action 1, and conduct a follow-up analysis after a few weeks for evidence of further amplification. This level of *Legionella* represents little concern, but the number of organisms detected indicates that the system is a potential amplifier for *Legionella.*
3	Implement Action 2, and conduct a review of premises for direct and indirect bioaerosol contact with occupants and health risk status of people who may come in contact with the bioaerosols. Depending on the results of the premises review, action related to cleaning and/or biocide treatment of the equipment may be indicated. This level of *Legionella* represents a low but increased level of concern.
4	Implement Action 3. Then, cleaning and/or biocide treatment of the equipment is indicated. This level of *Legionella* represents a moderately high level of concern, since it is approaching levels that may cause outbreaks. It is uncommon for samples to contain numbers of *Legionella* at this level.
5	Immediate cleaning and/or biocide treatment of the equipment is clearly indicated. Conduct post treatment analysis to ensure effectiveness of the corrective action. The level of *Legionella* represents a high level of concern, since it poses the potential for causing an outbreak. It is uncommon for samples to contain numbers of *Legionella* at this level.

Excerpted from *Reducing Risks Associated with Legionella Bacteria in Building Water Systems.*[22]

Other Pathogenic Bacteria

This section excludes *Legionella pneumophila.* These other "airborne pathogenic bacteria" of potential concern to the environmental professional are not as well studied yet still deserve attention.

Disease and Occurrence of Prominent Airborne Pathogenic Bacteria

As they may not be identified during air or diagnostic sampling, disease or the chance presence of pathogenic bacteria in a routine bioaerosol air sample may be the only means for suspecting its presence in a given environment. In alphabetical order, typical disease and occurrence of airborne pathogenic bacteria frequently posing a concern are herein discussed.

***Bacillus anthracis*:**[35,36] The genera *Bacillus* are spore-forming, rod-shaped bacteria which require oxygen to grow. Although there are numerous species, only the *Bacillus anthracis* is known to be lethal to man. Many species have a commercial use in insect control in the agriculture and forest industry. Although *Bacillus anthracis* is the only deadly *Bacillus*, other forms may be opportunistic. Some species of *Bacillus* and several genus of *Clostridium* share the commonality of being the only pathogenic, spore-forming bacteria. However, *Bacillus anthracis* may cause disease through airborne transmission.

Mostly a disease of lower animals, *Bacillus anthracis* is transmissible to man via the skin, alimentary tract, and respiratory tract. Disease in man is typically an occupational disease associated with butchers, shepherds, herdsmen, and handlers of hides, hair, and fleece. During World War I, anthrax-contaminated articles (e.g., shaving brushes) from Asia and South America provided a source for infection. Presently, reports of anthrax disease come from Haiti and Zimbabwai.[37] The primary concern is for airborne exposures which result in pulmonary anthrax.

Pulmonary anthrax is due to inhalation of the microorganisms from the air. Although uncommon, it is the most dangerous. It occurs occupationally among those who handle/sort wool and fleece where the spores are floating in the air from the infected material. It is characterized by symptoms of pneumonia which frequently becomes fatal septicemia. The dose-response is low. Research indicates that only a few inhaled spores are required to produce disease. Immunization is possible, and infections are treated with antibiotics.

During World War II, *Bacillus anthracis* was researched heavily due to its airborne pathogenicity and resistance of the spores to drying. An effort was made to develop drug-resistant strains which is not clear as to their success.

They did, however, find that the spores can remain viable in soil for as long as sixty years.

Bacillus anthracis is also a hazard to textile workers working with imported animal hair. The primary concern may be airborne exposures, but the most likely source of infection is through the skin.

The genus *Bacillus* are Gram-positive rods, measuring 4.5 to 10 microns in length by 1 to 1.25 microns wide. They grow well on all common nutrient media, most rapidly between the temperatures of 41° C and 43° C in aerobic conditions. Spores can be destroyed by any of the following methods:

- Boil for ten minutes.
- Heat at 140° C for three hours under dry conditions.
- Disinfect with 3 percent hydrogen peroxide for one hour or with 4 percent potassium permanganate for fifteen minutes.

Corynebacterium diphtheriae:[38] Only one species of *Corynebacterium* is an airborne pathogen. That is *Corynebacterium diphtheriae* the cause of diphtheria.

Diphtheria results from airborne exposures to the bacterium. The *Corynebacterium diphtheriae* enters the body via the respiratory route, lodging in the throat and tonsils. Death may result from suffocation by blockage of the air passages and by tissue destruction from the toxin evolved.

Occasionally, it has affected the larynx, resulting in membranous croup, or the nasal passages, causing membranous rhinitis. Diphtheria infections of the conjunctiva and of the middle ear are less common, and cutaneous or wound diphtheria is only occasionally observed. However, wound diphtheria may be serious resulting in a systemic infection. Systemic infections can severely affect the kidneys, heart, and nerves. Primary infection of the lungs and diphtheritic meningitis have been observed on rare occasion. Thus, most of the concerns for the diphtheria bacillus would be in hospital environments.

Corynebacterium diphtheriae are generally described as Gram-positive rods with irregular staining responses. The ends of each rod are swollen and respond greatest to staining, a characteristic unique to the genus. The bacteria measure 1 to 6 microns in length and 0.3 to 0.8 microns in width.

Mycobacterium tuberculosis:[39,40] *Mycobacterium tuberculosis* has three varieties (e.g., *hominis*, *bovis*, and *avian*). The *Mycobacterium tuberculosis* variety *hominis* is highly infectious via airborne transmission of the disease. Not only are there different varieties, but there are different strains.

Resistance to drug treatment occurs *in vitro* and *in vivo.* In the infectious stage, drug resistance is more common with the *Mycobacterium tuberculosis* than other pathogenic bacteria because of the extended, lengthy antibiotic treatments required to fight the disease. Their persistence allows greater chances for mutated strains to develop.

Tuberculosis is one of the most important communicable diseases in the world, affecting more than fifty million people. In the United States, over fifty thousand new cases are reported annually, and ten thousand resultant deaths occur per year. However, disease does not always develop in all those who have been exposed. There is a possible dose-response relationship.

Potential transmission occurs in public places and hospitals, typically by airborne droplets of contaminated sputum. Other modes of infection (e.g., genitourinary tract, conjunctiva of the eyes, skin, and alimentary tract) are less common.

Bacteria-laden droplets and dust particles are inhaled, settle in the lungs, and grow. In an individual with low resistance, an infection occurs. Extensive destruction and ulceration of the lung tissue progress to other parts of the body, predominately the spleen, liver, and kidneys. If not controlled by antibiotics, death may result.

The tuberculosis bacteria are "acid-fast" rods, measuring 2 to 4 microns in length by 0.3 to 1.5 microns wide. Specialized enriched media and aerobic conditions are required for growth and isolation of the bacillus. The optimum temperature for growth of the mammalian varieties is 37° C with a range of 30° C to 42° C.

Various Genera of *Pseudomonas*:[41,42] *Pseudomonas* includes over thirty species which are found in water, soil, and compost. Of the many species, there are only a few pathogenic types.

The most widely known pathogenic form is *Pseudomonas aeruginosa* (also referred to as *Pseudomonas pyocyanea*). They grow readily on all ordinary culture mediums and most rapidly between the temperatures of 30° C and 37° C. Although they typically require oxygen, some will grow in anaerobic conditions. As they produce a bright blue-green color which diffuses into its substrate, a blue-green color has, at times, been observed on substrates (e.g., surgical dressings) where the bacteria grew.

Pseudomonas aeruginosa frequently occurs as a secondary invader of infected tissues or tissues which have been traumatized by operation. Children and immune suppressed patients are particularly susceptible to infection. Oftentimes, infections are associated with the urinary and respiratory tracts. Abscesses may develop in different parts of the body, especially the middle ear. Although rare, it may also cause endocarditis, pneumonia, meningitis, and systemic pyocyanic infection. The latter generally occurs in patients with severe burn wounds where the bacterium enters and spreads throughout the body, causing endotoxic shock and focal necrosis of the skin and internal organs. As they do not respond well to antibiotics, systemic infections are almost always fatal. For this reason, the primary area of concern for this bacterium is in hospitals.

It should be noted that bacteriostatic detergents (e.g., cationic detergents) provide a growth environment for certain Gram-negative bacteria (e.g., *Pseudomonas aeruginosa*). Many of these soaps are use in skin antiseptics, mouthwashes, contact lens solutions, and disinfectants for hospitals.

Pseudomonas aeruginosa is also an insect pathogen and has consequently been considered for use as an insecticide. Its limited shelf life and potential to cause disease in humans has been a deterrent from its use as a pest control agent.

Pseudomonas pseudomallei infections may be asymptomatic or result in acute, toxic pneumonia or overwhelming septicemia. The organism has been isolated from moist soil, market fruits and vegetables, and well and surface waters in Southeast Asia. *Pseudomonas maltophilia* is suspect as well. However, over the past few years, the latter suspect has undergone numerous name changes and is now referred to as *Stenotrophomonal maltophili*. It is no longer classified as genus *Pseudomonas*.

The *Pseudomonas* bacterium are Gram-negative rods, measuring 1.5 to 3 microns in length by 0.5 microns wide. They grow well on all common nutrient media and grow most rapidly between the temperatures of 30° C and 37° C in aerobic conditions. Once airborne, the bacterium loses its viability with drying. So, exposures must occur shortly after the bacteria have become dispersed into the air.

Sampling and Analytical Methodologies

If pathogenic bacteria are suspect, the sampling equipment of choice for low to moderate levels (e.g., less than $10{,}000/m^3$) are the culture plate impactors (e.g., Burkhard sampler and Andersen impactor). In situations involving potentially higher numbers, liquid impingers are preferred, primarily due to their ability to dilute the solution and plate from several diluents. Notwithstanding surface samples, filter air sampling is easier but may theoretically be used only for the spore-forming *Bacilli anthracus* for which analysis is not available commercially. *Mycobacterium tuberculosis* is also of questionable feasibility.

Although air sampling for *Mycobacterium tuberculosis* has been attempted by some environmental professionals, it is not recommended by most microbial laboratories. Where the concern involves this pathogen, alternate surface sampling may provide more reliable results. Otherwise, a negative air sample can not be considered conclusive as to the nonpresence of the pathogen.

In the laboratory, samples which are not already on a culture medium are plated. Filter surfaces are washed, the suspended material plated on nutrient media. Impinger solutions are also plated.

Most pathogenic bacteria will grow on tryptic soy agar or a nutrient blood agar, the same nutrients used to sample for total bacteria during indoor air quality studies. Although the nutrient media of choice is blood agar, the microbial laboratory chosen to perform the analysis should be consulted for their preferred media.

Incubation should take place at 35° C for most pathogenic bacteria. Pathogenic bacteria tend to grow best at elevated temperatures, and the higher temperatures restrict growth of most environmental bacteria which could overrun a sample.

With a few exceptions, colonies should be counted within one to two days. One such exception is *Mycobacterium tuberculosis* which require up to two weeks for colonies to become visible.

Where shipping is necessary, a count may take place upon receipt of a well-insulated shipment of culture plates. Impinger samples should be plated prior to shipping to avoid growth within the collection solution during transportation.

Interpretation of Results

Experience and careful consideration of comparative and/or diagnostic samples are important for developing sound conclusions. The sampling methodologies alone may damage some of the otherwise viable, unprotected bacteria. Thus, the mere identification of specific microbes in an indoor environment, particularly in hospitals, may be cause for remediation.

The outdoor environmental contribution of a pathogenic bacterium may also be important in evaluating the findings and devising a remediation plan. Where the levels are as high or higher outdoors, indoor exposure controls may be more difficult or, in some cases, not required.

Where air sampling is performed for *Mycobacterium tuberculosis*, positive findings of any level should be acted upon. However, negative results may not have any meaning. Negative findings should not be relied upon where false negatives, particularly involving drug-resistant strains, can lead to a lethal, false sense of security.

Limitations of Sampling and Analytical Methodologies

Pathogenic bacteria sampling is limited by the lack of experience/published research information and further development of methodologies, and, once again, no guidance standards. As clinical laboratories infrequently care to become involved with environmental sampling for pathogens, environmental, microbial laboratories capable of performing most analyses may lack experience and/or the necessary equipment for identifying some of the less commonly requested pathogens.

BACTERIAL ENDOTOXINS[43,44]

Endotoxins are toxic components of the lipopolysaccharide, cellular membrane of Gram-negative bacteria. Although the bacteria may be rendered nonviable, the endotoxin retains its toxicity, even through extremely high temperatures (up to 110° C) which may not normally be exceeded in an autoclave.[43] Their impact on the individual is dose related, and airborne levels have been reported from a variety of work environments.

The most commonly reported cases have involved processing of vegetable fibers, fecal material in agriculture, and human waste treatment plants. Most worker exposures occur in cotton gins/mills, swine confinement buildings, poultry houses, grain storage facilities, sewage treatment facilities, and wood chip processing/saw mills. Other areas include the aerosolizing of machining fluids in metal processing,[45] aerosolized contaminated water supplies in hospitals,[46] and "humidified" office buildings.[47] See Table 4.6 for a recap and amplification on the environmental settings in which endotoxins have been identified at levels significant enough to cause illness.

Table 4.6 Occupational Environments Where Endotoxins Have Been Identified and Known to Pose a Problem[47]

Cotton Gins/Mills
Swine Confinements
Grain Storage, Handling, and Processing Facilities
Poultry Barns
Sewage Treatment and Processing Facilities
Wood Chipping Operations
Saw Mills
Flax Mills
Machine Shops
Fiberglass Manufacturing [48]

Symptoms typically involve elevated temperatures. In the cotton industry, this is referred to as "mill fever." Onset of fever is followed by malaise, respiratory distress (e.g., coughing, shortness of breath, and acute air flow obstruction), diarrhea, vomiting, hemorrhagic shock, tissue necrosis, and death. However, the latter, more serious symptoms are rare.

Some researchers have also been able to demonstrate acute changes in the respiratory FEV_1 and suggest repeated exposures may cause a syndrome similar to chronic bronchitis.[49] Other investigators feel that endotoxins add to the virulence of a parasite, enhancing the perils of disease. Once again, hospitals provide a special niche for such occurrences. Paradoxically, on the same note, it is also thought that the endotoxins which threaten one's health can enhance the body's immune system by challenging pathogenic bacteria, viral infections, and cancer.[48]

Sampling and Analytical Methodology

Bulk sampling of humidifier reservoirs and other potential endotoxin sources is considered the most reliable means of sampling for endotoxins. All samples should be collected in sterilized glassware and analyzed promptly, before microbial and subsequent endotoxin amplification can occur. Sterilization of all glassware and associated laboratory equipment should be baked at 210° C for one hour to destroy most pre-existing endotoxin contaminants. Plastic items should also have been sterilized (e.g., ethylene oxide) or sonication in endotoxin-free one percent triethylamine. As they may adsorb large amounts of lipopolysaccharides, supplies made of polypropylene plastics should be avoided. Polystyrene is preferred!

Air sampling has been performed with limited success. Attempts have been made by dust/particle collection techniques. The methods of collection have included total dust samplers, cascade impactors (size separators), vertical elutriators (used for respirable dust sampling in the cotton industry), cyclones (used for personal respirable dust sampling), and midget impingers (personal samples which collect particles greater than 1 micron in size). As endotoxins act primarily in the lower region of the lungs (which mostly involves particles of less than 5 microns), "respirable dust samplers" are preferred. For those considering the possibility of viable sampling for bacteria to get a gauge as to endotoxin levels, endotoxin and viable bacteria samples have not been found to correlate.[50]

Sample collection flow rates have varied, based upon the type of sampler. They have been as low as 2 or 4.5 liters per minute for total dust to as high as 7.4 liters per minute with the vertical elutriator. Although not specified, one study suggests that in order to detect down to 14 picograms per cubic meter, a minimum of 6.5 cubic meters of air should be sampled.[51] Another study sampled at 2 liters per minute for the duration of a workshift, for a volume of 960 liters.[52] It is quite apparent that laboratories vary in their analytical methods and, therefore, their detection capabilities. For this reason, the laboratory should be consulted prior to sampling.

Filter media are also worthy of mention. In one study, recovery of endotoxin from four different types of filters was compared to a "no filter" control.

The recovery rates, as reflected in Table 4.7, were extremely poor (e.g., polyvinyl chloride membrane filter) to moderate (e.g., mixed cellulose ester membrane filter).[51] In a later publication, the same investigator who performed the filter media testing proposed the use of "0.4-µm polycarbonate filters" in a standard protocol for sampling endotoxins.[53]

Table 4.7 Recovery Efficiencies of Endotoxin from Filter Media

Filter Type	Recovery (ng of NP-1 activity/ ng LPS added)	Percent Recovered as Compared to the Controls (%)
Polyvinyl chloride (5.0-µm pore membrane)	0.064	10.8
Polyflon (woven Teflon)	0.206	34.4
Teflon membrane (1.0-µm pore membrane)	0.229	38.2
Cellulose mixed ester (0.45-µm pore membrane)	0.252	42.1
Lipopolysaccharide (LPS) control (no filter)	0.599	

NP-1 = Reference endotoxin
Excerpted from Endotoxin Measurement: Aerosol Sampling and Application of a New Limulus Method.[51]

Upon selection of sampling media, all supplies should be pre-cleaned and sterilized, including the filters and cassettes. Cassettes can be cleaned by sonication in endotoxin-free 1 percent triethylamine, and all other supplies which might come in contact with the sample (including the laboratory implements) should be autoclaved at 210° C for 1 hour.[52] As ability to perform these processes is generally not within the realm of the environmental professional, supplies (including filters) are best provided by the laboratory.

After collection, the samples and a field blank (unsampled filter) should be sealed in airtight plastic enclosures, placed in a cold storage container, and sent immediately to the laboratory. Do not freeze the samples.[51] Each of the samples should be analyzed at the same time, by the same method, within the same laboratory.

In the 1940s, endotoxin was measured by injecting rabbits and monitoring the increases in body temperature. The sensitivity was 100 picograms, but rabbits and preparations were inconsistent.[55] Referred to as the Pyrogen Test, this method remained in place until the discovery of the unique effects endotoxin had on the blood cells of the horseshoe crab.

In 1977, the U.S. Food and Drug Administration licensed the more sensitive Limulus Amebocyte Lysate (LAL) Assay as an alternative to the Pyrogen Test. An extract is made from amebocytes of the *Limulus polyphemus* (or horseshoe crab). In the presence of endotoxins, clotting occurs. There is a relationship between the amount of endotoxin present and the rate/amount of clotting.

Samples are extracted and tested according to one of four principal methodologies which are based upon blood cell clotting. Extraction methods vary as much as the analytical methods. In a proposed standard method, however, the recommended procedure is to "extract samples in 0.05 M potassium phosphate, 0.01 percent triethylamine, pH 7.5, using bath sonication."

Analytical methods include the gel-clot test, calorimetric test, chromogenic test, and Kinetic-Turbidimetric Limulus Assay with Resistant-parallel-line Estimates (KLARE) Test. Due to its precision, sensitivity, resistance to interferences, internal validation of estimates, and ability to provide quantitative as well as qualitative information, the KLARE test is preferred by researchers and there is an attempt to standardize the use of this test as well.[56] However, despite all the studies and research, the choice of method is going to be laboratory dependent.

Interpretation of Results

Results are reported in terms of equivalent weight per volume (e.g., ng/ml or ng/m^3), equivalent mass (e.g., ng/mg) or potency endotoxin units per volume (e.g., EU/m^3). The U.S. Reference Standard EC-5 for endotoxin has a conversion factor of 10 EU per ng. of substance.

With the results in hand, interpretation is precarious at best. There have been no standardized means for interpreting endotoxin results. Yet, a number deserves some form of processing.

In 1986, one researcher observed a threshold of 10 nanograms per cubic meter (ng/m^3) for acute reductions in pulmonary function tests and of 50 to 100 ng/m^3 for fever amongst exposed workers. These conclusions were based upon dust samples collected by vertical elutriator with subsequent analysis by LAL. As the same analytical method performed the same by two separate laboratories can yield differing results, the present feasibility of setting a threshold limit is like setting a limit on taxes. In the present arena, it doesn't work!

Many of the more prominent researchers and commercial laboratories tend toward comparative sampling between complaint and non complaint areas. Relative differences should shift simultaneously along with variances created by the different approaches. When sampling for endotoxins in a metal fabrication shop around the cutting fluids, a comparative sample may be collected in an office. When sampling in a turkey processing plant, a comparative sample may be collected in the administrative area.

According to the ACGIH Committee on Bioaerosols (1989), sample results which exceed the background blank by a multiple of one hundred or more, coupled with endotoxic symptoms, should be considered excessive. Excessive levels should culminate in remedial activities (e.g., humidifier cleaning).

Limitations of Sampling and Analytical Methodologies

Identification of an analytical laboratory may be difficult, and interpretation of the findings is subject to type of sampling methodologies utilized. Not only are the methods variable, but the filter media are variable. Field and laboratory handling of the sampling/analytical tools and equipment requires proper planning and preparation. Sterile techniques during sampling involve additional care which is not normally a concern for the environmental professional. Comparisons of intralaboratory results and varying analytical times are not recommended. Consistency in approach and analytical method will provide more consistency in interpretation.

PATHOGENIC PROTOZOA[57,58]

Typically larger than the bacteria and mold spores, protozoa are unicellular microorganisms which are free-living and thrive in water. They may be located in damp soil, mud, drainage ditches, puddles, ponds, rivers, and oceans. Those which represent an occupational and environmental concern are the amoebae, measuring in size from 8 to 20 microns in diameter.

Amoeba move by flowing pseudopodia, or false feet, and usually live in fresh water. They are naked, or unprotected, during vegetation. Otherwise, they secrete a protective shell. They may be found in home humidifiers where they have been known to cause an allergic reaction called humidifier fever. The genera *Naegleria* and *Acanthanoeba* have been implicated in building-related illness, yet most amoeba-related disease is associated with waste treatment plants.

In the absence of water, the *Naegleria* can encyst to protect itself. These spherical cysts are 9 to 12 millimeters in diameter. The seriously parasitic amoebae, which are not typically found in fresh water, are not generally a concern.

Sampling and Analytical Methodology

Sampling equipment may be the Litton sampler, two in-series all-glass impingers, or sieve plate impactors. High volume air samples should be collected close to the probable aerosolization source, avoiding other potential contaminant sources (e.g., not immediately under running water). Equipment should also be sterilized.

The Litton sampling tube should be cleaned, using seventy percent ethanol and submerged in two separate distilled water rinses for 30 to 60 minutes each. The last rinse should be collected and analyzed for background amoeba contamination of the tubing.

The all-glass impingers should be cleaned, and the final rinse water should be analyzed for background contamination. The final in-series samples (consisting of 150-milliliters of water) should be combined and analyzed collectively.

At the laboratory, each sample solution should be plated in five different amounts onto nonnutrient agar plates with the common bacteria *Escherichia coli* and incubated between 43° C and 45° C. Pathogenic amoebae will grow more rapidly than non pathogenic types in these conditions.

The amounts of solution per sample to be plated should be 0.01-, 0.1-, 1, 10-, and 100-milliliters. They are treated differently. The 10-milliliter sample should be centrifuged at 500x gravity for 15 minutes, and the product plated. The 100-milliliter of sample should be filtered through a 1.2 micron pore sized, cellulose filter which must be quartered or halved and inverted in an agar plate. The smaller aliquots should be plated directly onto separate plates. Assure all plates have been properly labeled.

Incubation may take up to seven days at 45° C (or at 35° C for some of the less heat tolerant species), and then a series of more complex manipulation techniques are performed. The resultant suspension requires an additional three hours of time-lapsed examinations for concentration determination of thermophilic *Naegleria* species. Speciation requires an additional two weeks *in vivo* mouse pathogenicity studies.

Interpretation of Results

Disease, coupled with reservoir source identification, is diagnostic and does not require quantitation. The existence of these heat-loving amoebae in a reservoir requires immediate remediation with an oxidizing biocide (e.g., sodium hypochlorite or hydrogen peroxide).

Limitations of Sampling and Analytical Methodologies

This method is time-consuming and labor intensive where time may be important. It is expensive and requires special experienced laboratory manipulation.

Alternate immunoassay techniques are quick, easy, and inexpensive. They are discussed in Chapter 6.

VIRUSES[59]

Viruses, the smallest living organisms, are obligate intracellular parasites. They are subdivided into animal, plant, and bacterial parasites. The size of animal viruses ranges from twenty to 300 nanameters.

Viruses are host specific and become invasive only when specific host organs become accessible. Host entry may be through any of a number of mechanisms, yet most enter through the respiratory tract.

A line of defense (e.g., nasal hairs and mucous secretions) must be passed. Then, the virus must be transported (e.g., through the blood supply) to its target cell preference(s). Once the target cells have been identified, the virus penetrates the cell wall barrier and takes command of the cell's replicating mechanism. The virus is reproduced within six to forty-eight hours. A protective structure is built around each replicated virus, and the progeny viron either destroys the host cell or forms a bud which allows it to pursue other host cells.

Animal viruses are usually recognized by the diseases they cause (e.g., AIDS). The greatest concerns with aerosolized animal viruses are influenza, measles, chicken pox, and some colds. Virulence is influenced by the following:

- Specific type of virus
- Concentration in an aerosol
- Aerosol particle size
- Individual susceptibility

Indoor contamination occurs in residences, offices, laboratories, hospitals, and animal confinement areas. Outdoor exposures occur around livestock, sewage treatment plants, caves, and water sources. Environmental factors affecting virus survivability are relative humidity, temperature, wind, ultraviolet radiation, season, and atmospheric pollutants.

Amplification of viruses does not occur without a host. Hence, increased numbers will not occur in water or organic substrates of air handlers. The air handlers will merely serve as a means of conveyance. Acids, extreme temperatures (e.g., less than minus 20° F), and drying may, however, damage or destroy most viruses. This information is important for an exposure limiting consideration for prevention and control of viruses and for handling samples.

Sampling is typically neither recommended nor requested. The presence of disease is generally tracked epidemiologically to the source or sources, and sampling is only performed where verification is requested. Where the sampling methodologies are not restrictive (any microbial sampling procedure is possible), analysis should be performed by an experienced clinical laboratory. Due to the complexity of the viral analysis as well as its infrequent use, the environmental professional should consult with the laboratory for details on handling, turnaround time, and interpretation of results.

DIAGNOSTIC SAMPLING METHODOLOGIES

Oftentimes, it becomes necessary to identify the source of amplified microbials. Either remediation of a suspected source has failed to rectify a problem, or the environmental professional chooses to confirm suspect sources at the time of the initial sample taking, possibly after obtaining the initial air sample results and prior to making recommendations.

Swipe and bulk samples can be used to diagnose suspect sources (e.g., medical equipment and drip pan water in the air handling system) of elevated levels of fungi. One technique is as follows:

- Identify the area to be sampled (e.g., four square inches of a surface or the entire surface area of a sterile trachea tube)
- Open and expose a sterile swab
- Swipe the surface of concern
- Streak a culture plate (chosen on the basis of the type of microbe sought) with a zig-zag motion
- Tape and label the sample plate

Some prefer using sterile water on the end of the sterile swab to serve as a more efficient collection device while others choose not to factor another potentially contaminated material into the equation. In either case, it is advisable to take blanks as well. At a minimum, open the swab and streak a culture plate. If sterile water is used with the swab, a water treated swab should be streaked on a second plate.

Bulk samples should also be taken under sterile conditions. If a liquid is collected, it should be collected by use of a sterile transfer device (e.g., sterile pipette). If the material is solid and requires handling, it should be handled by sterilized material (e.g., sterile gloves) and equipment.

A bulk sample may, however, be obviously contaminated by visible microbial growth (e.g., *Penecillium* on gypsum board). In such cases, an attempt should be made at least to avoid touching the suspect material, and a sterile container should be used for transfer to the laboratory. Plastic zip-lock baggies can provide a convenient means for collecting and transferring large items.

REFERENCES

1 Bailey, M. Robert, Ph.D. and E.G. Scott. *Diagnostic Microbiology.* C.V. Mosby Company, St. Louis, Missouri, 3rd Edition, 1970.

2 Conant, Norman F., Ph.D., et. al. *Manual of Clinical Mycology.* W.B. Saunders Company, Philadelphia, 3rd Edition, 1958.

3 U.S. Department of Health, Education, and Welfare. *Occupational Diseases: A Guide to Their Recognition*. U.S. Government Printing Office, Washington, D.C., Revised Edition, June 1977.

4 Rhame, Frank S., M.D. Endemic Nosocomial Filamentous Fungal Disease: A Proposed Structure for Conceptualizing and Studying the Environmental Hazard. *Infection Control*. 7(2):126 (1986).

5 Krasinski, Keith, M.D., et. al. Nosocomial Fungal Infection During Hospital Renovation. *Infection Control*. 6(7):278-82 (1985).

6 Weems, J. John, M.D., et. al. Construction Activity: An Independent Risk Factor for Invasive Aspergillosis and Zygomycosis in Patients with Hematologic Malignancy. *Infection Control*. 8(2):71-5 (1987).

7 Cross, Alan S., M.D. Nosocomial Aspergillosis: An Increasing Problem. *Journal of Nosocomial Infection*. 4(2):6-9 (1985).

8 Mascola, J.R. and L.S. Rickman. Infectious causes of carpal tunnel syndrome: case report and review. *Rev Infect Disease*. 13(5):911-7 (1991).

9 Kilburn, C.D. and D.S. McKinsey. Recurrent massive pleural effusion due to pleural, pericardial, and epicardial fibrosis in histoplasmosis. *Chest*. 100(6):1715-7 (1991).

10 Center for Disease Control. Update on Coccidioidomycosis in California. *Morbidity and Mortality Weekly Report*. 43(23):421-3 (1994).

11 Chen, G.H., et. al. Case Records of the Massachusetts General Hospital—Weekly Clinicopathological Exerciser. *New England Journal of Medicine*. 330(7):490-6 (1994).

12 Paya, C.V. Fungal Infections in solid-organ transplantation. *Clinical Infectious Diseases*. 16(5):677-88 (1993).

13 Pfeiffer, T.J. and D.H. Ellis Environmental isolation of Cryptococcus neoformans var. gattii from Eucalyptus tereticornis. *Journal of Medical and Veterinary Mycology*. 30(5):407-8 (1992).

14 Pleil, J.D. Demonstration of a Valveless Injection System for Whole Air Analysis of Polar VOCs. *Proc. 1991 Int. Symp. Measurement of Toxic and Related Air Pollutants*. Air and Waste Management Association, Pittsburgh, Pennsylvania, 1991.

15 Bailey, M. Robert, Ph.D. and Elvyn G. Scott. *Diagnostic Microbiology*. C.V. Mosby Company, St. Louis, Missouri, 3rd Edition, 1970.

16 Burge, Harriet A. *Bioaerosols*. Lewis Publishers, Boca Raton, Florida, 1995. p. 259.

17 Morris, George K., Ph.D. and Brian G. Shelton M.P.H. *A Suggested Air Sampling Strategy for Microorganisms in Office Settings*. [Technical bulletin] PathCon Laboratories, Norcross, Georgia, 1994.

18 ACGIH Committee on Bioaerosols. *Guidelines for the Assessment of Bioaerosols in the Indoor Air Environment—Fungi.* ACGIH, Cincinnati, Ohio, 1989.

19 ACGIH Committee on Bioaerosols. *Guidelines for the Assessment of Bioaerosols in the Indoor Air Environment—Mycotoxins.* ACGIH, Cincinnati, Ohio, 1989.

20 Cox, Christopher S., and Christopher M. Wathes. Bioaerosols Handbood. CRC/Lewis Publishers, Boca Raton, Florida, 1995. p.375.

21 Burge, Harriet A. *Bioaerosols.* Lewis Publishers, Boca Raton, Florida, 1995. p. 90.

22 Cox, Christopher S., and Christopher M. Wathes. Bioaerosols Handbood. CRC/Lewis Publishers, Boca Raton, Florida, 1995. pp. 375-6.

23 Ibid. p. 376.

24 Aflatoxin–Nature's most potent carcinogen. [Handout] Neogen Corporation, Lansing, Michigan, 1996.

25 Stewart, F.S. *Bacteriology and Immunology for Students of Medicine.* Williams and Wilkins Company, Baltimore, Maryland, 9th Edition, 1968.

26 Morris, George K., Ph.D. and Brian G. Shelton, M.P.H. *Legionella in Environmental Samples: Hazard Analysis and Suggested Remedial Actions.* [Technical Bulletin 1.4] PathCon Laboratories, Norcross, Georgia, 1995.

27 Brock, Thomas D. and Michael T. Madigan. *Biology of Microorganisms.* Prentice Hall, Englewood Cliffs, New Jersey, 6th Edition, 1991. pp. 518-520.

28 Burge, Harriet A. *Bioaerosols*--Legionella *Ecology.* Lewis Publishers, Boca Raton, Florida, 1995. pp. 49-76.

29 Shelton, Brian G., et. al. *Reducing Risks Associated with Legionella Bacteria in Building Water Systems.* Prevention and Control of Legionellosis. [Technical Bulletin 2.4] PathCon Laboratories, Norcross, Georgia, 1995.

30 Legionella from misting in grocery store. New York Times, January 11. 139:A1(N) 1990.

31 Hlady, W. Gary, et. al. Outbreak of Legionnaire's Disease Linked to a Decorative Fountain by Molecular Epidemiology. *American Journal of Epidemiology.* 138(8):555-62 (1993).

32 Alcamo, I. Edward, Ph.D. *Fundamentals of Microbiology.* 3rd Edition, Benjamin/Cummings Publishing Company, Inc., Redwood City, California. p. 243.

33 Gorman, George W., James M. Barbaree, James C. Feeley, et. al. *Procedures for the Recovery of Legionella from the Environment.* [Bulletin] CDC, Atlanta, Georgia, November 1992.

34 Sheldon, Brian G. Social Security Building incident where settlement is alledged to have been quite expensive. [Oral communication] PathCon Laboratory, Norcross, Georgia, July 1995.

35 Atlas, Ronald M. and Richard Bartha. *Microbial Ecology: Fundamentals and Applications.* Benjamin/Cummings Publishing Company, Menlo Park, California, 1987. p. 476.

36 Burrows, William, Ph.D. *Textbook of Microbiology.* W.B. Saunders Company, Philadelphia, 1968. pp. 614-619.

37 Morris, George K., Ph.D. Exposures to *Bacillus anthracis.* [Oral communication] PathCon Laboratories, Norcross, Georgia, November, 1995.

38 Brock, Thomas D. and Michael T. Madigan. *Biology of Microorganisms.* Prentice Hall, Englewood Cliffs, New Jersey, 6th Edition, 1991. pp. 513-514.

39 ACGIH Committee on Bioaerosols. *Guidelines for the Assessment of Bioaerosols in the Indoor Air Environment—Bacteria.* ACGIH, Cincinnati, Ohio, 1989.

40 Stewart, F.S. *Bacteriology and Immunology for Students of Medicine.* Williams and Wilkins Company, Baltimore, Maryland, 9th Edition, 1968. pp. 357-60.

41 Ibid. pp. 278-280

42 Ibid. pp. 282-283.

43 ACGIH Committee on Bioaerosols. *Guidelines for the Assessment of Bioaerosols in the Indoor Air Environment—Endotoxins.* ACGIH, Cincinnati, Ohio, 1989.

44 Burge, Harriet A. *Bioaerosols.* Lewis Publishers, Boca Raton, Florida, 1995. p. 78.

45 Gordon, Terry. Acute Respiratory Effects of Endotoxin Contaminated Machining Fluid Aerosols in Guinea Pigs. *Fundamental and Applied Toxicology.* 19:117-23 (1992).

46 Burrell, Robert. Human Responses to Bacterial Endotoxin. *Circulartory Shock.* 43:137-53 (1994).

47 Jacobs, Robert R. Airborne Endotoxins: An Association with Occupational Lung Disease. *Applied Industrial Hygiene.* 4(2):50 (1989).

48 Jacobs, Robert R. Airborne Endotoxins: An Association with Occupational Lung Disease. *Applied Industrial Hygiene.* 4(2):52 (1989).

49 Reynolds, Stephen J. and Donald K. Milton. Comparison of Methods for Analysis of Airborne Endotoxin. *Applied Occupational Environmental Hygiene.* 8(9):761-7.

50 Rietschel, Ernst Theodor, and Helmut Brade, *Scientific American.* 267(2):55-61 (1992).

51 Milton, Donald, et. al. Endotoxin Measurement: Aerosol Sampling and Application of a New Limulus Method. *American Industrial Hygiene Association Journal.* 51(6):331-7.
52 Reynolds, Stephen J. and Donald K. Milton. Comparison of Methods for Analysis of Airborne Endotoxin. *Applied Occupational Environmental Hygiene.* 8(9):762.
53 Burge, Harriet A. *Bioaerosols.* Lewis Publishers, Boca Raton, Florida, 1995. p. 83.
54 Burge, Harriet A. *Bioaerosols-Endotoxin.* Lewis Publishers, Boca Raton, Florida, 1995. pp. 77-86.
55 Milton, Donald, et. al. Endotoxin Measurement: Aerosol Sampling and Application of a New Limulus Method. *American Industrial Hygiene Association Journal.* 51(6):333.
56 Burge, Harriet A. *Bioaerosols.* Lewis Publishers, Boca Raton, Florida, 1995. p. 82.
57 ACGIH Committee on Bioaerosols. *Guidelines for the Assessment of Bioaerosols in the Indoor Air Environment—Protozoa.* ACGIH, Cincinnati, Ohio, 1989.
58 Burge, Harriet A. *Bioaerosols.* Lewis Publishers, Boca Raton, Florida, 1995. pp. 122-123.
59 ACGIH Committee on Bioaerosols. *Guidelines for the Assessment of Bioaerosols in the Indoor Air Environment. Viruses.* ACGIH, Cincinnati, Ohio, 1989.

Chapter 5

ANIMAL ALLERGENIC DUST

The neglected partner in allergenic complicity with pollen and mold spores is animal allergens in "house dust." Only within the past ten years have clinical studies revealed a strong relationship between levels of animal allergens in dust and allergy symptoms. Technology has evolved. Methods have been refined, and immunoassay technology comes into the limelight.

Detection and quantitation of a wide range of antigenic biological and non biological substances are now possible through "immunoassay analytical methods." Allergenic substances which are processed include proteins, glycoproteins, hormones, peptides, chemical haptens, and drugs. Of particular interest to the environmental professional, researchers have developed immunoassays for animal allergens, predominantly those derived from mites, cats, cockroaches, and rodents. Methods have also been developed for certain species of fungi (e.g., *Aspergillus flavis*) and for latex.

An immunoassay involves identification of the antigens by creating antibodies for the express purpose of tagging specific materials. Quantitation is based on the antibody-antigen complexes. Thus, immunoassay analyses are highly specific and quantifiable.

The sampling procedure is simple and inexpensive. Yet, the sampling strategy and results interpretation require a thorough understanding of the process.

In the past, the medical community has performed the sampling, but most of the sampling has been diagnostic, involving expensive clinical tests performed on the distressed sufferer. Where allergies appear widespread in an office building or other problematic indoor air environment, the perplexed facilities manager or home owner seeks assistance from the environmental professional. Diagnostic tests on all the building occupants can be expensive and time consuming. In such instances, dust sampling is by far the more feasible alternative.

Yet, without the benefit of clinical studies, extensive allergy complaints may pose a medley of possibilities. Some of the allergy sufferers know the specific antigens which cause their individual reactions. Known allergens may assist in narrowing the possibilities, where dust sampling may be performed in order to:

- Define allergen levels in residences of asthma patients.
- Identify areas or sources of elevated levels of allergen(s).
- Determine the effectiveness of allergenic-dust control measures

Most of the information provided within this chapter is intended to aid in the search for the more common allergens and expound on those which may be overlooked in isolated instances. The animal allergens are more widely understood and an evolving issue of concern in the environmental field.

ANIMAL ALLERGENS

Animal proteins are high molecular weight, complex molecules which can illicit an allergic reaction. Wherein the environmental professional is concerned with airborne exposures, the allergens must be present in large quantities and small enough to become airborne. Typically, those animal allergens which are more commonly encountered are parts-and-pieces of an insect or mammal. Those which receive the greatest attention and are frequently studied are dust mites, dog/cat dander, and cockroach body parts. The probability of elevated numbers of these allergens is considerable in most indoor air environments.

Mites/Spiders[1,2]

Mites are small to microscopic-sized, generally parasitic arachnids with four pairs of legs in their adult stage and little or no differentiation of the body parts. Many cause allergic rhinitis, human dermatitis, and general allergic reactions. They differ in habitat and associations and are broadly categorized by their associations. See Table 5.1 for a breakdown of the most commonly cited allergenic mite types.

"Storage mites" are usually of the genera *Lepidoglyphus* and *Tyrophagus*. They rely on decaying vegetation as a food source and are normally associated with agricultural environments. They have been identified as causing allergic rhinitis in dairy farmers. Thus, allergy-causing storage mite exposures are limited more to the outdoor environments where there is decaying vegetation than to the indoor air environment. Decaying vegetation is a requirement for their presence.

"Itch mites" and mange mites may cause a dermatitis and have occasionally been implicated with house dust mites. Their main food sources are cheese, dried meats, flour, and seeds. They damage and contaminate these commodities while having a means of transportation through humans handling of the food products. The result is "grocer's itch," or miller's itch.

Some attack man and other animals directly, burrowing into their skin. These are the ones which hikers and hunters often encounter in wooded areas.

Others cause mange which results in itching and hair loss. Mange mites typically attack domestic animals and are generally visible to the naked eye. Yet, following a heavy infestation, dead or alive, their bodies may still serve as antigens.

Table 5.1 Allergenic Mites

	Class: Arachnida	
	Order: Acari (Acarina)	
	Suborder: Psoroptoidea	
STORAGE MITES		
	Family: Acaridae, Glycophagidae, and Blomia	
	Genus/species:	*Acarus siro*
		Glycophagus domesticus
		Lepidophagus destructor
		Tyrophagus putrescentiae
		Blomia tropicalis
ITCH MITE		
	Family: Sarcoptidae	
	Genus/species:	*Sarcoptes scabiei*
DUST MITES		
	Family: Pyroglyphidae	
	Genus/species:	*Dermatophagoides pteronyssinus*
		Dermatophagoides farinae
		Dermatophagoides microceras
		Euroglyphus maynei

Excerpted from *Allergy Basics for IAQ Investigations.*[3]

"House dust mites" have undergone considerable study as they are not only allergenic, but they typically are found indoors. They bask in warm, moist, dark environments. Ideal temperatures are between 70 and 80° F. They thrive where the relative humidity is in excess of 65 percent, and they hide from sunlight.

Sites where they tend to commune are places that have sluffed epithelial cells (which tend to retain moisture), such as beds, upholstered furnishings, and carpets. The average human will lose as much as five grams of epithelial skin cells per week. Wherever these epithelial cells can be found, the mites have a source of food.

In the United States, dust mite infestations and allergies tend to be seasonal with a preference for the warmer, humid months (e.g., summer).

Whereas tropical climates may provide a perpetual, unrestricted habitat for blissful invaders all year round, the pesky little critters predominate mostly between June and August.[4]

The house dust mites are between 250 and 500 microns in size, barely visible by the naked eye and frequently overlooked. The allergenic portion of these mites is thought to be the body parts-and-pieces and their fecal material which is 10 to 35 microns in diameter. The body pieces are considerably smaller than 250 microns and not identifiable by microscopic analysis. As a matter of course, the size has a bearing on the airborne allergens. If greater than 40 microns, airborne substances will settle out within 20 to 30 minutes. Thus, inhalation of the material is most likely to those components of the dust which are smallest and in areas often disturbed.

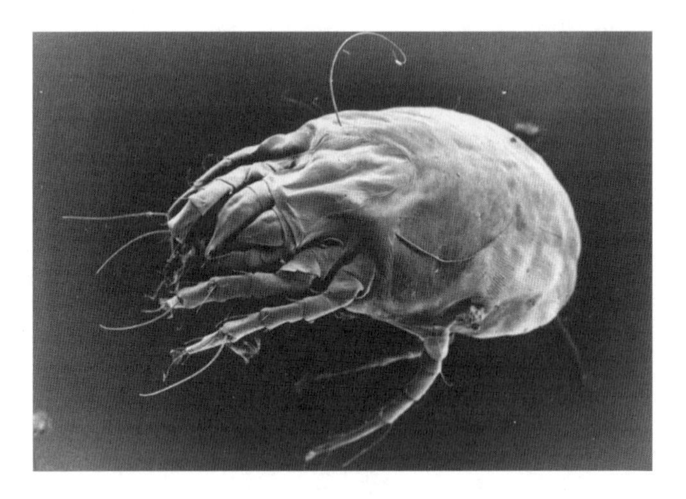

Figure 5.1 The dust mite is a commonly used representation of allergens. (Courtesy of Vespa Laboratories, an ALK-Abello Company, Spring Mills, PA)

Areas likely to be disturbed are situation dependent. Commonly involved activities which might stir up dust include, but are not be limited to, the following:

- During and shortly after vacuuming
- Considerable activity on and disturbance of upholstered furniture
- Considerable activity on and disturbance of carpeting
- When making a bed and fluffing pillows
- Sleeping on contaminated bedding and/or upholstered furniture

As for the actual allergens, several mite-associated proteins are implicated, and studies have been predominately on three species in the genera

Dermatophagoides which is common in North America and Europe. To a lesser extent, the genera *Euroglyphus* and *Blomia* have been studied as well, but they are more commonly encountered in Central and South America.

For the purpose of allergen testing using immunoassay techniques, the dust mite allergens are genus and species specific, and each of the species has as many as forty different proteins which could cause an allergic reaction. Where the specific allergenic proteins have been identified, they are referred to by "Group." See Table 5.2 for the most commonly implicated allergen types and associated allergenic proteins.

Where a group of allergenic proteins have not been identified, a homogeneous mix is referred to as polyclonal. A polyclonal assay involves multiple antigens from the same life form.

Table 5.2 Allergenic Dust Mites and Immunoassay Test Groupings

Type	Allergenic Proteins Group I	Group II	Group III
Dermatophagoides farinae	*Der f* I	*Der f* II	*Der f* III
Dermatophagoides pteronyssinus	*Der p* I	*Der p* II	*Der p* III
Dermatophagoides microceras	*Der m* I	—	—
Euroglyphus maynei	*Eur m* I	—	—

Booklice

Oftentimes, the layman will refer to paper mites as being the source of a problem, possibly because the individual associates their allergies with the mounds of paper they work with and street hearsay. The reference to "paper mites" is a red herring, a fictitious contrivance of the news media. Entomologists frown in a desperate attempt to track these illusive pests under the heading of mites. While some entomologists will confess ignorance, others will speculate that the reference is more likely to that of storage mites which, at times, are associated with cellulose, or paper products. Another consideration is that of "booklice" which are neither mites nor lice, but insects.

Booklice belong to the order Psocoptera. These are small, soft-bodied insects with three pair of legs and measuring less than 1/4 inch in length. They may or may not have wings. They have been reported as causing allergic symptoms in places with large amounts of paper. They feed on molds, fungi, cereals, pollen, and dead insects. Their preferred habitat is moist areas and humid environments, and they rarely cause damage to the spaces they occupy. They are, however, a frequent nuisance to allergy sufferers.

Figure 5.2 Photomicrograph of an "unconfirmed" booklice, a component of office dust, lifted by tape. Its approximate size is 500 microns, and this image was observed, in its entirety, under 100x magnification. Immunoassay testing was for cockroach allergens only which were found to be excessive. Dust mite allergens were low, but the method was specific for dust mites, not booklice.

Cockroaches and Other Insects[3]

Insects are typically visible, have three pairs of legs in their adult stage, and possess three distinct body regions. They are, therefore, easy to identify, and a large indoor insect population rarely remains unnoticed. The most common is the ever-present cockroach.

Of the 55 species of cockroaches that inhabit the continental United States, less than ten are indoor residents. Of these, the most common, particularly in southeastern United States areas, are the larger American cockroach (*Periplaneta americana*) and the smaller German cockroach (*Blatella germanica*). As the larger ones consume the smaller, more prolific ones, they do not tend to cohabit within the same residence. It should be visually apparent as to which species one is dealing with. See Table 5.3 for the allergenic groups.

Table 5.3 Allergenic Cockroach Material and Immunoassay Test Groupings

	Allergenic Proteins	
	Group I	Group II
Blatella germanica	*Bla g* I	*Bla g* II
Periplaneta americana	*Per a* I	—

Recent studies, however, suggest that the cockroaches secrete their allergen onto their bodies and other surfaces in their environment. Thus, examination of allergenic material may or may not disclose the presence of associated debris and fecal material. The only means of confirming the presence of cockroach allergens is through immunoassay analysis of suspect dust.

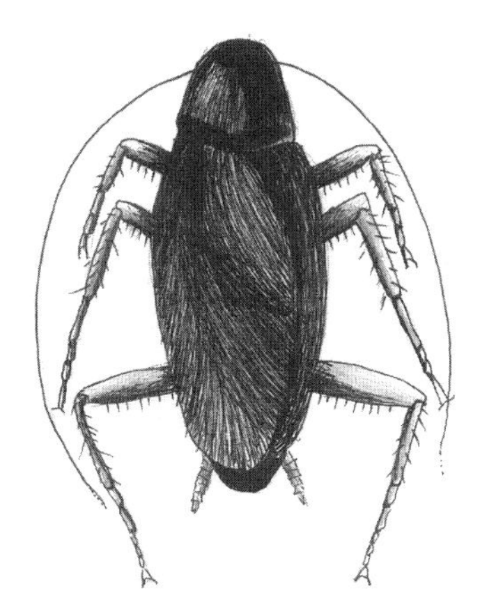

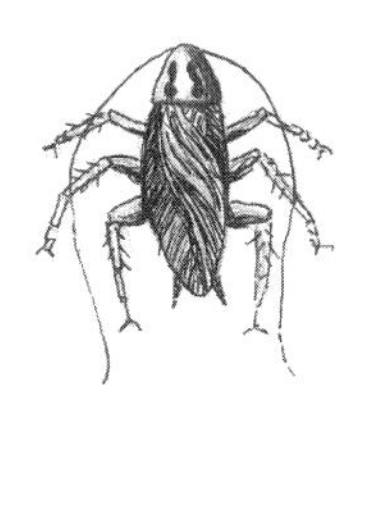

Figure 5.3 (left) The American cockroach (*Periplaneta americana*) and (right) the German cockroach (*Blatella germanica*).

There has been considerable, unconfirmed speculation as to the source of the cockroach allergens. Some considerations are as follows:

- Saliva
- Body parts-and-pieces
- Egg shells
- Fecal particles

Cockroaches are able to adapt to low ambient humidity, yet actively seek a source of water. For this reason, indoor cockroaches are most likely found around water pipes, pet water bowls, evaporative areas around refrigerators, leaking faucets, and wet carpeting. Although in most cases their presence is readily apparent, cockroach allergens have been measurable in up to 15 percent of homes which had no visible clues that they might be present. Keep in mind that they do not have to be alive for the allergens to promote a reaction, and the source of cockroach allergens is still unclear.

Although there have been numerous allergen studies performed of the ever-enduring, ever-present cockroach, many other insects have been implicated as well. Though most other insects are generally found in outdoor environments, body parts-and-pieces may be conveyed indoors. They may attach to clothing. They may enter open windows and doors in search of food or light (e.g., June bugs). There may be air movement from the outside to enclosed air spaces indoors. Indoor accumulations of insect carcasses are common. The indoor environment may become a repository of debris.

Outdoor workers are exposed, at times, to insect fragments and debris at levels in excess of ragweed pollen. Insects which are suspect of causing allergies include the following:[3]

- Crickets
- Houseflies and fruit flies
- Waterfleas
- Bed bugs
- Mayflies
- Aphids
- Honey bees
- Bean weevils
- Some species of moths
- Butterflies
- Silkworms
- Caddis flies
- Chiromomid midgets

Some occupations, due to their associations, harbor potential exposures for insect infestations. Some examples may be found in Table 5.4.

Table 5.4 Occupational Exposures to Insects

Entomologists: locusts, crickets, flies
Grain mill workers: beetles, grain weevils
Loggers/lumber mill workers: Tussock moth
Fishermen/bait handlers: meal worms, maggots
Poultry workers: Northern fowl mites
Bakers: storage mites, grain weevils
Small animal handlers: fleas
Honey-packing plant workers: honey bee dust
Pet food processors: Chironomids

Domestic Animals[6]

Most allergic substances which are associated with domestic animals are predominately exposure problems in environments where the animals reside. The obvious is overstated. Homes, kennels, pet shops, and laboratories are locations where the allergens are most likely to be found. However, office environments should not be excluded from consideration. Although rare, elevated exposures have been reported where the source has not be obvious. The target domestic animals are cats and dogs.

Cats

Although cats are maintained in 28 percent of all American households, only two percent of the U.S. population has allergies to them. Interestingly, those who are allergic to cat allergens may never have lived with cats.

The source of allergenic material may be any of a number of feline-associated materials, and tests are performed for *Felis domesticus* (*Fel d* I). There has been considerable speculation as to the actual chemistry of the allergen, but most researchers speculated that the allergen is somehow transferred, picked-up, and/or concentrated by saliva. When the cat grooms itself by licking, the allergen is spread or transferred to the hair and epithelial cells. The following in an abbreviated list of known and suspect sources/transfer vehicles:

- Saliva
- Sebaceous glands
- Hair
- Epithelial cells
- Epidermis

The sex and type of cat are factors which contribute to environmental levels of cat allergen. Male cats shed more allergen than female cats. Patients symptoms vary in severity, depending upon the type of cat they are exposed to (e.g., a domestic cat versus a Persian short-tail).

The cat allergens are carried on particles less than 2.5 microns in diameter. At this size, after becoming airborne, they will remain suspended in an undisturbed environment for hours. The ease with which these allergens become airborne is the reason for apparent excesses in sensitivity.

Even though there are excessive cat allergens in a household where one resides, all indoor environments have detectable levels. In houses with cats, there is typically in excess of 10 micrograms per gram (μg/g) of *Fel d* I in the dust, and levels have been reported as high as 7,000 μg/g.[7] The allergen may also be transferred by clothing and other articles from a high exposure environment to otherwise cat-free environments. Houses which have never had cats may have levels of less than 1 μg/g in the dust, but levels in excess of this should not be surprising.

Dogs

Dogs are maintained in an estimated 43 percent of American homes, and one regional study indicated that as many as 17 percent of the population was allergic to dog allergens. There have been twenty-eight different allergens found to be associated with allergic symptoms. The specific antigen to which most patients react is designated as *Can f* I. *Can f* I consists of an extract of the following dog-associated materials:

- Hair
- Dander
- Saliva

Most environments where dogs are found have in excess of 120 μg/gm. of *Can f* I in the dust. Homes without dogs typically have less than 10 μg/gm. The size of the allergen-carrying material or contaminated particles is unknown.

Rodents[8,9]

Exposures to mouse and rat allergens are typically associated with animal research laboratory vivarian and indoor spaces (e.g., homes and office buildings) infested by rodents. Of the approximately 35,000 workers in the United States exposed to rodent allergens in animal research laboratories or breeding facilities, over 20 percent of the workers experience allergic symptoms.

Rodent infestations may deposit allergens unbeknownst to building occupants. Awareness of the potential opens another door for search and disclosure of possibilities. Complaints of a urine-like odor should alert suspicion.

Two allergens have been associated with mice. One of the mouse allergens, referred to as Ag1, is related to the mouse urine and is designated *Mus m* I. *Mus m* I is produced by the liver and salivary glands, excreted in the form of urine and saliva. As it is associated with testosterone, *Mus m* 1 is excreted predominately by male mice. Other factors affecting quantity are strain and age. The other mouse allergen, referred to as Ag3, has been detected in hair follicles (e.g., fur and dander extracts). This is designated Ag3.

Two allergens have also been associated with rats. However, they are both associated with the rat urine. These allergens, referred to as Ag4 and Ag13, are designated *Rat n* I.

Although typically associated with particles of 10 microns in size or less, airborne exposures to rodent urine rarely occur unless contaminated bedding is disturbed, and elevated humidity has been reported to diminish airborne exposures. Most exposures occur where rodents are maintained in large numbers (e.g., laboratory environments). The amount of material that may become airborne is generally related to the type of litter and bedding. The allergen is generally released into the air during cage-cleaning activities. In one laboratory animal cage-cleaning study, the airborne levels were reported between 19 and 310 nanograms per cubic meter of sample (ng/m^3). During quiet times, when there were no disturbances, the levels dropped to around 1.5 to 9.7 ng/m^3. Rodent allergens may also be deposited on ceiling tiles and in carpeting. Disturbances of contaminated areas may result in airborne releases of material as yet not identified to be present in a given environment (e.g., rodent infestations). There is no published information regarding reported airborne or dust levels of rodent allergens.

Farm Animals[10]

Along with the arthropods, pollen, mold spores, and bacteria, farmers and farm workers are potentially exposed to farm animal allergens. The more prominent allergenic exposures are attributed to cows, horses, and pigs.

Cow dander and urine, designated *Bos d* II, have been reported to cause allergic rhinitis in dairy farmers. Airborne levels have been reported as high as 19.8 μg/m^3.

Horse allergens (*Equ c* I, *Equ c* II, and *Equ c* III) are very potent. Exposures may occur occupationally or to pleasure horseback riders. The horse allergens are related to hair, dander, and epithelial cells.

Pig allergens are rarely reported to be a problem and are considered weak antigens. The allergenic material has, however, been identified. Swine workers have been found to have antibodies against swine dander, epithelium, and urine. Airborne levels have been reported up to 300 μg/m^3. Yet, there appear to be minimal complaints and concerns for swine allergies.

Other Animals[10]

Rabbit dander and guinea pig urinary proteins/saliva have been reported to cause allergies. Both are found in homes, pet stores, and laboratory facilities.

Even rarer are exposures to bat guano and reindeer epithelial cells. Bat droppings accumulate inside roof attics and cave dwellings. Asthma-like symptoms are generally reported in association with workers exposed to bats in indoor working environments.

Reindeer epithelial cells which are associated with leather processing are also known to cause allergic reactions. Airborne exposure levels have been reported in a workshop at concentrations of 0.1 to 3.9 μg/m^3.

OCCURRENCE OF ANIMAL ALLERGENS

Farm, laboratory, and pet environments are easy marks. The source of allergies is direct and readily apparent to the allergy sufferer when symptoms worsen in their presence. The greater the number of animals, the greater the potential for elevated exposures. Whereas an individual may not at one time have been sensitive to a given allergen, an extreme dose may later predispose them to developing symptoms at lower exposure levels in the future. Most of these allergy sufferers know what it is they are allergic to. A small percentage of the population, however, seems to be sensitive not only to the typical allergens, but to just about everything.

In office environments with no apparent sources of animal allergens, the latter more allergen-sensitive individuals will be the first to start complaining. It has been estimated that these ultrasensitive individuals constitute only about 4 percent of the population. Yet, as levels of an allergen increase, the numbers impacted increase as well. With more complaints comes greater concern for locating a source. As animal allergens are possible contributors to a given environmental invasion, a means for identifying and quantifying their presence is made available through a well thought-out strategy, collecting dust samples, and analyzing the collected material by immunochemisty.

SAMPLING STRATEGY

A well thought-out strategy is vital for identifying a problem and obtaining meaningful results. If the environment is an office building and large numbers of people are impacted, the problem areas must be clearly identified.

Questionnaires should be filled out by all those in the area of concern as well as an area where there are no complaints of allergy-like symptoms. Identify "known" problem areas and non problem areas. Develop associations. Attempt to limit the possibilities. A screening tool for rodents, using ultraviolet light, may also be added to the list of considerations. The method is discussed in the next section of this chapter.

Other than bacterial and mold spores, the most typical allergenic materials found in office buildings are dust mites and cockroach allergens. Cat and dog allergens are more common in homes but may be transferred to office environments. All other allergens previously mentioned are rare occurrences in office/industrial environments or are occupationally related.

Some building occupants may have had allergy testing (e.g., skin tests or allergy blood tests) performed and know what allergens to which they are allergic. A few of the more common blood tests available through physicians include the following:[11]

- Plant pollen
- Fungi/molds
- Thermophilic actinomycetes
- Storage mites
- Animal dander
- Isocyanates
- Formaldehyde
- Gums/adhesives
- Anhydrides

The greatest dilemma to the environmental professional is that of locating areas most likely to be source origins and include only these in the sample, not the disassociated areas as well. For instance, if complaints generally arise in a given area where there is a lot of disturbance of the carpeting and dust mites are suspect, the area where the traffic passes should be sampled to include as much of the known problem material as possible. Including areas under desks and along walls may or may not be the source origins. Each falls within a different source type (e.g., carpeting) and functional grouping.

Sample sites should be selected based on suspect source types. These may include, but not be limited to, the following locations:

- Carpeting
- Upholstered furniture
- On top of ceiling tiles
- Ducting in an air handling unit

Functional grouping of sample sites is the most difficult to identify. The impacting function may or may not be obvious. In most cases, it will not be obvious, and several functional areas will require sampling. Grouping may include, but not be limited to, the following:

- Dusting shelves
- Vacuuming the carpeting
- Excessive traffic
- Maintenance involving removal of ceiling tiles
- Air handler activity

The number of samples taken will depend upon the environmental professional's assessment of the situation. This will vary in a case by case situation, depending upon the number of possibilities ascertained to be a potential problem source. Then, too, sampling of a non problem area may be desirable for comparative purposes.

During data/information gathering, clarify whether a suspect carpet was recently shampooed or vacuumed. Maybe there has been a recent infestation of cockroaches or rodents. Even where the vermin have since been exterminated, their body parts-and-pieces may be the exposure allergens. In locating possible sample sites where these parts-and-pieces may have been deposited, the environmental professional should also consider the function of that location as well. For instance, rodents may eat through air supply ducting and deposit leavings. Allergenic material deposited in an air plenum will pose a greater potential for occupant exposures than the same material deposited in the corner of a room where there is no foot traffic or air movement.

Record the sample area size and exact location. Although the area size may not be relevant (samples are analyzed by weight comparisons), this additional information may be useful at some future date, and some professionals do standardize the sample size (e.g., one cubic meter of area). You, once again, are not obligated to do the same.

Although there is no set protocol, sample sites should be clearly identified. If floor plans are available, indicate the limits of each site and assign a sample number to the area. Otherwise, describe in detail the specific location(s) (e.g., on top of the ceiling tile above the copy machine) with their perceived function (e.g., frequent above ceiling work necessitates disturbance of ceiling tile at this location).

SCREENING FOR RODENTS[12]

Public health food inspectors screen for the presence of rodents in a food processing plant by using ultraviolet light. Under ultraviolet light, urine will fluoresce bluish white to a yellow white. Fresh stains fluoresce blue while the older stains shift to a more yellow color. Rodents tend to urinate while in motion, thus leaving a characteristic droplet trail. Rodent hairs will also fluoresce bluish white and can be easily identified in areas where they hang out (e.g., food storage areas).

This characteristic of urine allows visual inspection for tale-tall signs in dimly lighted areas. The darker the area under investigation, the more clearly visible will be the fluorescent stains.

SAMPLING PROCEDURES

Due to the complexity and lack of clinical comparison studies, sampling is typically performed on settled dust. Even though air sampling can as easily be performed, airborne levels vary considerably, because they are based on dust generating activities which are in progress during the sampling period. Depending upon the airborne dust levels, the required sample air volume may be in excess of 5,000 liters. Large air volumes, in turn, embrace extended sampling times and/or environmental air sampling devices which are capable of drawing 100 to 1,000 liters per minute. If sampling times are extended, the activity which generates airborne dust may not be singularly represented in the sample. Then, environmental sampling equipment is expensive and cumbersome. There are no published procedures for "allergenic dust" air sampling, and researchers shy away from this approach. On the other hand, settled dust sampling is simple and clinical comparison studies have been performed.

Settled dust sampling is as easy as sucking suspect dust into a vacuum cleaner bag or as involved as using specialty sampling devices designed to perform allergen dust sampling. The following sample collection devices have been used successfully by environmental professionals and allergists:[13]

Standard vacuum cleaner with a filter bag—The filter bag is later detached and sent to the laboratory for analysis. Studies have indicated that allergenic material thought to be retained in the hose preceding the bag should not be a significant concern to the sampler. They did, however, find that high retention, high efficiency bags do retain material better than the regular bags. Losses may be as much as 30 to 50 percent with the low retention bags. These may be purchased at some grocery/discount stores.

Commercial high-efficiency particulate (HEPA) vacuum cleaner with a high efficiency filter bag—The filter bag is later detached and sent to the laboratory for analysis. This apparatus is more efficient for retaining particles smaller than 2.5 microns in diameter than most conventional vacuum cleaners. To avoid cross-contamination, assure the hose and all connectors between the intake wand and filter have been cleaned prior to each collection.

Specially designed vacuum cleaner with an external filter attachment—One design incorporates a small bag that is inserted between the hose and the wand connection. Another uses a molded plastic wand that contains a filter.

Air sampling pump with a polycarbonate membrane filter cassette—Samples are drawn through the filter with an air sampling pump in much the same fashion as a vacuum cleaner (disregarding the flow rate) and collecting the dust directly from the surface area in question.

A private laboratory performed comparison tests for some of the above sampling devices and concluded that there is a significant difference in their allergen dust recovery. They ranged from half the original sample dose to double. For this reason, the same method should be used consistently, and comparison sampling of problem and nonproblem areas is strongly indicated so as not to rely on threshold values only.[13]

Commercial laboratories emphasize "the principal consideration in sampling should be the quantity of material captured." Although some researchers propose sampling within a well defined, delineated area or a limited sample duration, the quantity of material collected may not provide a good representation of the environment or allow for a sufficient amount of collected dust to retain the desired sensitivity (in the picogram range). Those who define the area of surface coverage generally opt for one square meter. Others sample for a specified time period (e.g., two minutes), no matter what the substrate. They typically vacuum for a set period of time (e.g., 5 or 10 minutes), without regard to the area covered. Yet, in both instances, the amount of dust collected still relies on the amount of available dust. There is no such thing as too much collected dust. There may, however, not be enough.

Ideally, the environmental professional should be able to estimate the amount of dust collected and target a collection in terms of milligrams (with the volume dependent upon the density of the collected material). Some laboratories specify 200 milligrams. Others specify 500 milligrams. The latter allows for a certain fudge factor with plenty to spare.

Although not required, composite sampling is recommended where several samples are to be taken and the primary allergenic reservoir has not been identified. One study indicated that three composite samples taken from the same dwelling, each a week apart, gave similar results. There was, however, a noted difference between areas within the same dwelling and between dwellings. Thus, composite samples provide a relatively consistent estimate, and discrete samples were useful in finding specific reservoirs.[13,14] These samples may be taken during the collection process or involve a contribution of dust from each of the samples taken in areas known to be associated with airborne allergens. This approach can be useful for screening and minimizing the number of samples requiring analyses for all allergenic dust.

In brief, the most relevant consideration is locating the sample site. Isolate and identify a specific, suspect sample site (e.g., around an area known to be associated with allergic reactions) along with its function/activity (e.g., high foot-traffic area or bedding). Then, collect a sample (or a composite of several samples). Compare suspect problem sites with known non problem sites. If an allergen appears suspect, discrete area sampling and analyses will aid in the identification of reservoir(s). Comparison sampling, coupled with published thresholds, will result in manageable interpretations.

ANALYTICAL METHODS

For the purpose of extending the reader's knowledge into the realm of understanding human test results and their applicability to a site investigation, human testing methods are discussed in brief within this section along with sample analytical methods. They are both relevant in assessing suspect allergens.

Human Testing

Airborne exposures to animal antigens may result in allergic rhinitis, sinusitis, and asthma. Although dermatitis and urticaria have not been implicated with airborne exposures, where an allergic individual rubs up against dust laden with a given antigen, the skin may be impacted as well.

Allergenic individuals may be tested by any of a number of means, each associated with a different route of entry, means of exposure. The simplest and most frequently method is the direct skin test.

The skin test technique primarily allows for identification of direct contact allergens only. It will not provide for identification of airborne allergens. A suspect antigen or group of antigens are placed on the skin surface of the individual, usually a site on the arm, and a retentive barrier is placed over the material. The site is checked for redness and urticaria after 24-hours. A positive reaction indicates the individual is sensitive to all or one of the challenge allergens.

A less commonly used technique is that of blood sample analysis. Blood is extracted from the individual and analyzed by immunochemical techniques. This method is by far the less invasive, not challenging the individual allergy sufferer. The immunochemical techniques used are the same as those used for the dust samples and are discussed in a little more detail under "Allergenic Dust Testing."

The most ideal (yet impractical) technique is that of a direct bronchial challenge to the individual. This method provides a direct insult to the individual while they are restricted to a challenge chamber where the airborne allergen types and amounts may be controlled. Whereas one may not respond to the skin test, the bronchial insult chamber may elicit a respiratory reaction. The technique is mostly of the research mode and, if accessible, very expensive and time consuming.

Allergenic Dust Testing

Commercial laboratories currently offer routine testing for dust mite, cockroach, cat, and dog allergens. Although an occasional commercial lab may be willing to extend their limits, several research institutions have the materials necessary to test for the less common allergens (e.g., *Mus m* I). Immunoassay testing of allergenic dust is extremely sensitive and highly specific. This specificity is advantageous in most cases but can be a drawback where a similar, but not identical, allergen is suspect.

Sample preparation involves sieving the dust samples to separate all material less than 300 microns in size from the larger material. After it has been weighed, the smaller material is then extracted with a special buffering solution. An aliquot is taken of this extract and analyzed by any of a number of immunoassay approaches, involving single antigen-specific or multiple antigen quantitation.

A single specific protein, or antigen, is referred to as "monoclonal." An example of a monoclonal test antigen is *Fel d* I. Being related to and being the strongest (or most studied) allergenic component of a given species, the monoclonal antigens become the most commonly sought after test material. Yet, due to the extensive amount of attention given to a limited number of allergenic species, the specificity of these methods may restrict the possibilities.

Those species which have received the most of the attention are also those which are known to cause allergies and are prevalent in significant numbers. Figure 5.4 shows the results of a study involving school children and prevalence of allergies. As a child ages, allergies diminish until ages 25 to 34.[15] Thus, the prevalence of the various allergies is likely to be less in the adult population.

Multiple proteins from one species or several are referred to as "polyclonal." An example of a polyclonal test antigen is cat allergens. They have not been as extensively studied as the monoclonal antigens, and results seem to be less consistent and are more difficult to evaluate.

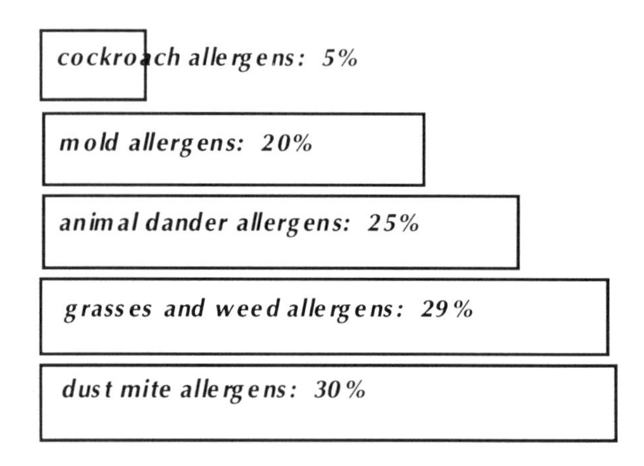

Figure 5.4 Estimated prevalence of allergies in school-age children and skin reactivity tests. Excerpted from *Indoor Allergens*.[16]

The identification of polyclonal antigens, however, may provide direction and assist the environmental professional in isolating the probable allergenic species in a dust sample. It serves well as a screening mechanism. More reliable, easier-to-use results may then be obtained through the monoclonal antigen tests. If, however, the screening fails to disclose any of the common allergens, all is not entirely lost.

An environment may be complicated by rare outbreaks or isolated occurrences. Elevated levels of fleas identified in the carpeting of a high foot traffic, medical reception rooms cannot readily be assayed. Dust which is known to be the source of allergies cannot readily be identified if the components are other than the more common species. Any of a number of scenarios may develop. In these instances, some laboratories will prepare test material from a given dust sample and test the blood of the allergy sufferers to confirm the allergenic potential of the material. The extracted dust is injected in a strain of mouse. If allergens are present, antibodies are formed. The newly created antibodies are then used to test for dust allergen in an allergy sufferer's blood. Sampling and interpretation require the assistance of a medical doctor.

INTERPRETATION OF RESULTS

A dose-response relationship is recognized with allergens as it is with toxic chemicals. A small percentage of the population will experience the effects at extremely low levels of exposure, levels not even noticeable to a majority of the population. These highly sensitive individuals are thought to comprise less than five percent of the work force. Although this is in line with chemical sensitivity numbers, the more sensitive individuals comprise a group of people who were: (1) predisposed at birth; and (2) exposed to low levels of specific antigens for a long duration. The remaining 95 percent of the population will be impacted as the exposure levels increase.

Ideally, "problem area composite samples" should be compared with "non problem area composite samples." Composite results may identify suspect allergens, and discrete samples will help to isolate the reservoir. Although most environmental professionals find great comfort in acceptable limits, comparisons with published thresholds should be accomplished with reserve.

Thresholds are in a constant state of flux, and recommended limits may vary from one laboratory to the next. Observations vary by region, and available species differ. Analytical results on the same sample may vary from one laboratory to the next. Laboratory findings and allergenic association tend to be variable, and these findings change within the same laboratory. For this reason, published thresholds should be used as a guide only!

For reference and review, some of the commonly accepted allergy-causing thresholds can be found in Table 5.5. Yet, they are not firm, hard-clad references. These limits are the result of observations made as to measured levels and resultant responses of typically nonatopic individuals. These reference values will not be applicable for those predisposed to allergies. In these atopic individuals (15 to 20 percent of the adult population), their thresholds may be one percent of that which will potentially impact the general population. It should also be noted that immunoassay reference thresholds for plant pollen, mold spores, and chemicals are not listed. This is due to the direct means available for determining airborne exposure levels for some and/or to the lack of data for reference thresholds.

Another publication sets different range limits, stating that their measurements are the most practical reference values available. They provide the following limits for Group I dust mite allergens:[20]

- Safe levels—less than 2 μg/gm.
- Levels that may sensitize atopic (genetically pre-disposed) individuals—2 to 10 μg/gm.
- Levels that may exacerbate previously sensitized individuals—greater than 10 μg/gm.

Table 5.5 Allergen Levels in Dust; Capable of Eliciting an Allergic Response in the General Population, Not Predisposed Genetically to Allergies

Allergens	Reference Thresholds (μg/gm.)
Dust mite allergens (polyclonal) *	15[17]
Der f I	2[17]
Der p I	2[17]
Group I	10[18]; 2[19]
Cockroach allergens **	
Bla g I	2[19]
Cat allergens **	
Fel d I	8[18]; 8[19]
Dog allergens **	
Can f I	—[20]

* *Dermatophagoides farinae* and *Dermatophagoides pteronyssinus.*
** Polyclonal antigens for which there are no published references.

It is thought that the Group I dust mite allergens are associated with fecal material which is relatively small in size as compared with the larger Group II allergens which are associated with the body parts. Thus, the Group I dust mite allergens are those most likely to be disturbed and become airborne. They are the most likely to be associated with allergy symptoms. The Group II body parts may complicate an evaluation where an individual is close to the settled allergenic material (e.g., sleeping on a contaminated pillow or lying on the floor). All conditions must be taken into consideration.[21]

Attempts have been made to report observed thresholds for airborne exposures to rodent allergens. One such attempt is summarized in Table 5.6. As these numbers are variable and not well studied, they should be used with reservations. At best, they may be considered reference guidelines. Dust air sampling should be performed through contributing input from the laboratory which will perform the analysis.

Table 5.6 Allergen Levels in Air Reported to Cause an Allergic Response

Thresholds	Observed Thresholds
Rodent Allergens	
Ag3	825 μg/m^3
Mus m I	59 μg/m^3
Rat n I	no reported observations

Excerpted from *Journal of Allergy and Clinical Immunology.* [22]

OTHER TYPES OF ALLERGENIC SUBSTANCES

Other types of allergenic substances include indoor/outdoor allergens and industry-related allergens. Indoor/outdoor allergens are mold spore allergens.

Mold spore dust sampling is rare. Easier, more commonly accepted approaches are available for analysis (as has been discussed in previous chapters), and there are limited guidelines for interpretation of results. In one report,[24] the researchers suggest that should the fungal spores and bacteria in dust exceed 10,000 colony forming units per gram of dust (CFU/gm) remediation may be indicated. Fungal concentrations on water-damaged materials (e.g., carpeting, gypsum board, or ceiling tiles) are excessive if they exceed 1,000 CFU/gm or 1,000 CFU/cm^2. However, the analysis for fungal spores relies on spore viability and is performed by the use of a culture media and Petri dishes. There are no immunoassay methods presently being used for mold spore quantitation in dust, but there are immunoassay methods to determine if an individual is allergic to "specific mold spores." History must dictate allergies, and the testing is performed for specific genera.

In certain industries, some of the more common allergenic proteins for which immunoassay methods may be performed have been identified in Table 5.7. Presently, however, processing of the samples may involve considerable expense and/or the aid of a research laboratory. There has been minimal or no dust sample analyses of these materials, and interpretation may be illusive.

Table 5.7 Allergenic Materials Which May Become Airborne

Industry	Type
Food processing	grain, flour, coffee bean, castor bean, egg, garlic, and mushroom dusts
Chemical/industrial	latex proteins, isocyanates, metals, resins, dyes, and drugs

Although each of the above allergens can be analyzed by immunoassay methodologies, many of the chemicals may be sampled using traditional industrial hygiene air and surface sampling methodologies. Once a laboratory has been identified to perform an analysis, the feasibility of performing air sampling for occupational allergens in the food processing industry will be based on one's willingness to collect comparative samples in nonproblem areas and comparing them to problem areas and/or with symptomology in order to establish an acceptable limit.

Then, also, the environmental professional may encounter a situation whereby dust from a grain elevator or adjacent food processing plant appears to be causing problems for the occupants of a building which is associated only by proximity to the food processing plant. In such cases, source dust samples in the immediate vicinity of a given building may be compared to dust collected indoors. Where symptoms indicate a probable exterior source, some of the occupants may also be tested with the aid of a medical doctor.

Of the chemical/industrial sources, latex protein analysis can be performed by immunoassay of a patient's blood. Drug dispersion sampling is often performed in-house by the pharmaceutical company laboratories, and the other chemicals can be sampled/analyzed as chemicals (not allergens) and compared with published reference limits for toxicity and/or sensitization levels.

Natural rubber latex proteins (produced from the sap of the *Hevea brasiliensis* trees) cause dermatitis to many of those who wear latex gloves, particularly to those in the medical professions where frequent use of the gloves is required. The frequent use and replacement of these gloves in an operating room are reputed to be the potential source of airborne latex proteins in the operating rooms. Of considerable concern, airborne latex proteins in an operating room, where the patient's system is accessible, may result in anaphylactic shock and possible death if the patient is already allergic to latex proteins. Another "surprise" use and source of airborne latex proteins has been reported in fireproofing material sprayed on structural members of buildings.

In one case, analysis of the fireproofing insulation was performed. The concentrations of latex protein were 1,000 to 2,000 nanograms per gram of material (ng/g). Samples of blood were taken from thirty-six occupants, and four tested positive. The dust tested positive (5 to 26 ng/100 cm^2), and the airborne concentrations ranged from nondetectable to 16 ng/m^3. The only confirmation of the latex allergens as being the cause of skin rashes and other allergic reactions was remediation which led to elimination of the symptoms. Although there were no controls, the latex-containing fireproofing was implicated as the probable source of building complaints.[25]

REFERENCES

1 Burge, Harriet A. *Bioaerosols.* Lewis Publishers, Boca Raton, Florida, 1995. pp. 134-137.

2 Borror, Donald J. and Dwight M. DeLong. *Introduction to the Study of Insects.* Holt, Rinehart and Winston, New York, Third Edition, 1971. pp. 634, 636-637.

3 Halsey, John F., Ph.D. and Mark Colwell, M.S. Allergy Basics for IAQ Investigations. (Handout) Professional Development Course, American Industrial Hygiene Conference & Exposition, May 20, 1995.

4 Lintner, Thomas J., Ph.D. and Kathy A Brame, B.S. The Effects of Season, Climate, and Air-Conditioning on the Prevalence of Dermatophagoides Mite Allergens in Household Dust. *Journal of Allergy and Clinical Immunology.* April 91(4):862-7 (1993).

5 Burge, Harriet A. *Bioaerosols.* Lewis Publishers, Boca Raton, Florida, 1995. pp. 138-139.

6 Ibid. pp. 151-153.

7 Lintner, Thomas J., Ph.D. Topics on Allergens. [Oral communication] Vespa Laboratories, Spring Mills, PA. October 1995.

8 Burge, Harriet A. Bioaerosols. Lewis Publishers, Boca Raton, Florida, 1995. p. 154.

9 Jones, Robert B. et. al. The Effect of Relative Humidity on Mouse Allergen Levels in an Environmentally Controlled Mouse Room. *American Industrial Hygiene Association Journal.* 56:398-401 (1995).

10 Burge, Harriet A. *Bioaerosols.* Lewis Publishers, Boca Raton, Florida, 1995. pp. 154-155.

11 Dr. Halsey. IBT Reference Laboratory, "Specializing in Environmental Allergen Testing." [Bulletin] IBT Reference Laboratory, Lenexa, Kansas. pp. 2-3.

12 Sylvania. Black Light Radiant Energy. [Engineering Bulletin 0-306]. Sylvania, Danvers, Massachusetts. (1996)

13 Halsey, John F., Ph.D. and Mark Colwell, M.S. Allergy Basics for IAQ Investigations. (Handout) Professional Development Course, American Industrial Hygiene Conference & Exposition, May 20, 1995.

14 Lintner, Thomas J., Ph.D. et. al. Sampling Dust from Human Dwellings to Estimate the Prevalence of Dermatophagoides Mite and Cat Allergens. *Aerobiologia.* April 10(1):23-30 (1994).

15 Barbee, R.A., et. al. Longitudinal changes in allergen skin test reactivity in a community populations sample. *Journal of Allergy and Clinical Immunology.* 79:16-24 (1987).

16 Pope, Andrew M., et. al. *Indoor Allergens: Assessing and Controlling Adverse Health Effects.* National Academy Press, Washington, D.C., 1993. p. 52.

17 Chapman, M.D., et. al. Monoclonal Immunoassays for Major Dust Mite Allergens, *Der p* 1 and *Der f* 1, and Quantitative Analysis of the Allergen Content of Mite and House Dust Extracts. *American Academy of Allergy and Immunology*. 80:184-94 (1987).

18 Trudeau, W.L. and E.Fernandez-Caldas. Identifying and Measuring Indoor Biologic Agents. *Journal of Allergy and Clinical Immunology*. August 94(2)2:393-400 (1994).

19 Platts-Mills, T.A.E. Allergen Standardization. *Journal of Allergy and Clinical Immunology*. 87:621 (1991).

20 Schou, Carsten, Ph.D., et. al. Assay for the Major Dog Allergen, *Can f* 1: Investigation of House Dust Samples and Commercial Dog Extracts. *Journal of Allergy and Clinical Immunology*. December 88(6):847-53 (1991).

21 Burge, Harriet A. *Bioaerosols*. Lewis Publishers, Boca Raton, Florida, 1995. p. 135.

22 Ibid. p. 136.

23 Twiggs, et. al. Immunochemical measurement of airborne mouse allergens in a laboratory animal facility. *Journal of Allergy and Clinical Immunology*. 69:522 (1982).

24 Trudeau, W.L. and E. Fernandez-Caldas. Identifying and measuring indoor biologic agents. *Journal of Allergy and Clinical Immunology*. 94:393-400 (1994).

25 McCarthy, J.F., K.M. Coghian and D.M. Shore. Latex Allergen Exposures from Fiberproofing Insulation. Presented paper under the heading of Indoor Air Quality at the American Industrial Hygiene Conference. May 22, 1995. p. 6.

Chapter 6

RAPID MICROBIAL FINGERPRINTING

Microbial fingerprinting involves the identification of microbes through their individual, unique genetic/chemical characteristics. For environmental use, the classic, time-tested methods are time-consuming and tedious, requiring as much as two to six weeks of extensive laboratory manipulation. Due to impending health concerns and the need for cost controls, rapid microbial fingerprinting methodologies have become desirable. As researchers refine the methodologies, environmental professionals are pressed to expand their horizons.

Until recently, the methods mentioned herein have been used mostly in research and clinical diagnostics. Perceived need and simplicity are the driving forces which are expanding their use. The methods of greatest interest to the environmental professional are the Polymerase Chain Reaction and the Microbial Identification System.

A commercial kit, based on the Polymerase Chain Reaction technology, has been developed for the detection of *Legionella* species, and kits are being developed for other pathogenic microbes as well. This approach involves semiquantitative detection of DNA from the generalized genus *Legionella* and from the more aggressive species *Legionella pneumophila* found in water sources. It requires only six hours turnaround time for limited identification and semiquantitation of the viable and nonviable *Legionella* organisms.

The Microbial Identification System is usable for pathogenic and nonpathogenic bacteria. Although it requires a bacterial colony formation (typically 24 to 48 hours), this method provides not only genus and species, but strain as well. Both methods are explained in depth

POLYMERASE CHAIN REACTION[1–4]

The Polymerase Chain Reaction is also referred to as the "PCR" process, Perkin-Elmer GeneAmp™ PCR, and EnviroAmp™ PCR process. The process is highly specific and extremely sensitive in its ability to differentiate biological systems through the uniqueness of their DNA/genetic blueprinting.

The process serves an important role in rapidly identifying microbial problem sources, particularly where an epidemic ensues without a known source of the infectious microbe(s). It serves as a rapid response mechanism to imminent illness.

Uses for the PCR Process

The PCR process has been proven particularly useful in responding to outbreaks of Legionnaires' disease. Recorded events where this approach has been used the most is on cruise ships, office buildings, and hot tubes. An outbreak results in symptoms which are diagnosed as Legionnaire's disease or Pontiac Fever. Once the existence of Legionnaires' disease is suspect, an investigation of the possible source(s) may be initiated while the disease is being clinically confirmed. Culturing takes days, sometimes weeks, while the PCR process can be accomplished within hours. Where lives may be at stake, the analytical time required by this process is of foremost consideration.

Outbreaks of *Legionella pneumophila* have been investigated using this method on cruise ships. Numerous samples can be collected and analyzed for identification and semiquantitation. Identification is specific for the genus *Legionella* and the species *Legionella pneumophila. Legionella pneumophila* has been implicated in 85 percent of all *Legionella* pneumonia illnesses. Of the other 30 known species, only 18 have been associated with disease, and the microbes are ubiquitous in almost all aqueous environments (e.g., lakes, rivers, ponds, cooling towers, and potable water supplies). For this reason, the amount present is relevant as well as its means of aerosolization.

Although there are no acceptable or nonacceptable levels for PCR environmental sampling, source samples (e.g., water) may be compared and appear to correlate to problem sources. In most cases, the investigators have found that the semiquantitative analytical designation of low, medium, and high levels is a fairly reliable indicator. These levels are not to be confused with levels found by the sample culturing techniques, discussed in Chapters 3 and 4. The PCR technique is sensitive to viable and nonviable microbes whereas culturing is only sensitive to viable, disease-causing microbes which are culturable on the media provide.

However, the PCR process identifies and quantitates "viable, disease-causing" and "nonviable, nondisease-causing" microbes as well. The correlation between total count of disease-causing and nondisease-causing versus disease-causing cultural counts is unclear. Yet, in the case of *Legionella* on cruise ships, implicated sources have had high PCR levels of total counts on *Legionella*, whereas other sample sources had nondetected or low levels as analyzed by the PCR process.

Efforts are also underway to expand the PCR process usage to other environmental arenas. The National Institute for Occupational Safety and Health

(NIOSH) is presently undergoing research on *Mycobacterium tuberculosis* for its use in environmental and occupational air sampling. Efforts are underway to evaluate the impactors (e.g., two-stage Andersen impactor) and the AGI sampler. Some consideration is also being given to the use of DNA-free, membrane filter sampling for *Mycobacterium tuberculosis* due to the hardy nature of the bacterium. Attempts to culture the viable microbes have been unsuccessful and of questionable efficacy. Other than the use of animal exposures studies within contaminated areas, the PCR process possesses the most promising possibilities for environmental exposure studies.[5]

Methods for evaluating *Giardia* and *Cryptosporidium* are also on the list of environmental exposures and clinical diagnostics to be researched under the PCR process. Clinical diagnostic methodologies must pass approval by the U.S. Food and Drug Administration prior to widespread usage. At this time, *Giardia* and *Crytosporidium* remain research topics. Clinical diagnostics have been approved for *Legionella* and *Clamydia.* As of the end of 1995, the PCR analyses for *Giardia* and *Mycobacterium tuberculosis* are not available commercially in either clinical or environmental laboratories.

Sampling Methodologies

Legionella is the only environmental concern for which analysis by the PCR process is available commercially. Thus, the sampling methods discussed herein are targeted toward taking water samples.

Sampling for *Legionella* is typically brought about by the need to locate a source or sources of a known disease-causing agent. Potential source reservoirs may involve as many as fifty locations, fifty samples. Practically speaking, the investigator will seek to collect a sufficient, representative sample while keeping the size of each sample down.

Although a 10 milliliter sample is the minimum requirement, a 100 milliliter sample may be desired where the anticipated levels are low or wherever composite sampling is performed (e.g., cooling tower water). Samples should be collected in precleaned containers which have been provided by the analytical laboratory. Latex gloves should be worn and changed each time another source is sampled to avoid cross-contamination from one sample to the next. Containers should be capped, sealed, and properly labeled. Be sure to number the sample and record specific locations where the sample was taken (e.g., composite of north, center, and south portion of drip pan in air handling unit #2), conditions (e.g., 110° F temperature in a hot tube), and special observations (e.g., apparent microbial growths). Samples should be shipped immediately to the laboratory as per their directives.

Although not required, blank samples of a known, uncontaminated water source may disclose preexisting contamination which occurred from either the sample containers, packaging and shipping procedures, or analytical processing. A blank will provide an additional measure for quality control.

Analytical Methodology

Deoxyribonucleic acid (DNA) is the blueprint from which all life forms get their information on how to build and maintain their life process. A single DNA strand consists of links, or nucleotides. Each nucleotide has three parts: a phosphate unit, a sugar unit (e.g., deoxyribose), and one of four bases (e.g., adenine, guanine, cytosine, and thymine). See Figure 6.1 for a simplified schematic of three DNA nucleotides chained to one another.

The four nucleotide bases make up command centers which reproduce themselves and instruct ribonucleic acid (RNA) in how to manufacture cellular products. The sequencing of the DNA bases is specialized and unique for each life form. The greater the differences in function, the greater the differences in "base sequencing." This is the basis for differentiating one life form from another using the PCR process.

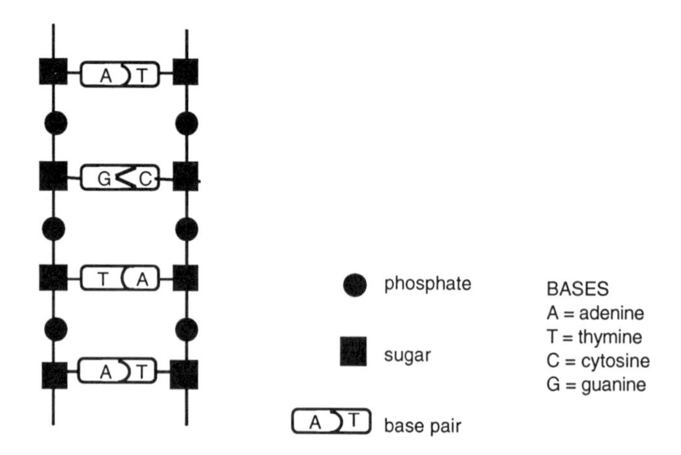

Figure 6.1 DNA chains consist of links of nucleotides, each with a sugar unit, a phosphate unit, and a base. Thousands of these nucleotides link together to form a DNA chain. The above examples are paired strands.

Each nucleotide base unit has its own complimentary base. Adenine only unites with thymine, and guanine only unites with cytosine. The reverse applies as well. A blueprint sequence may be any combination of the four bases. Although the sequences of the bases have yet to be uncoded for all life forms, many sequences have been determined and entered into a central gene repository for use by others.

Sequence variability occurs within the same species and diminishes with strains. Individual variability also exists to a lesser extent, particularly in lower life forms. Thus, each life form has similar and dissimilar base sequences. The more dissimilar the life form, the more dissimilar the base sequencing. In the PCR process, certain reproducible sequences are targeted. These target sequences are different from other closely related life forms and consistently the same within the life form sought (e.g., *Legionella*). Thus, there exist target sequences for genus, species, strains, and individuals.

The actual PCR process involves preparation, amplification, and detection. Preparation involves the release of DNA from its cellular bondage so their unique "target sequences" may be located. These target sequences do not include the entire stands of DNA. They are only a portion of that which is available, and the targets range in length from 100 to 10,000 base pairs. In bacteria, the DNA strands are as small as 1,000 base pairs. In man, they are as long as 10^9 pairs.[6]

The first step is to release the DNA strands. All cells in the sample are lysed, or broken open, thermally or chemically, in such a fashion that the typically more resilient DNA structures remain intact. This process is more difficult where the cell wall is more resistant to environmental influences (e.g., *Mycobacterium tuberculosis*). The approach differs by species.

Although the intent is to minimize DNA damage, some breakage of the chains is inevitable. For this reason, one organism cell per 10 ml sample may go undiscovered due to the cell preparation process.

The second step is amplification. The targets are small in size and number. The genus *Legionella* has four targets on the 5S rRNA gene, twenty-one to twenty-four bases per target. *Legionella pneumophila* has three targets on the *mip* gene, seventeen to twenty-three bases per target. Detection requires amplification.

The DNA sample is added to a PCR reaction tube which has the following:

- "Primers" with biotin markers—Primers are synthesized equal and opposite base sequences which are capable of recognizing the targets or duplicates and of duplicating their own kind. If a target differs by one just base in the sequence, it will not pair up with the primer.
- Special didoxynucleotides (which attach to growing primers with the aid of a polymerase)—The bases for these are those generally associated with RNA (where urasil is used in place of thymine).
- Heat stable DNA polymerase.
- Buffer.

In the reaction chamber, a cycle of reactions is started. Each cycle involves the following set of events:

- Denaturing—Unzipping the double stranded DNA chains into single strands.
- Annealing—Zipping the single DNA strands to their primers when there is a match.
- Polymerase extension—Adding more nucleotides to the primers in an order that is complementary to the single strand of DNA where the target sequence is found.

All the above reactions are performed in a special "thermal cycle" which rapidly cycles the temperature delivered to each reaction tube while it is being continuously mixed. Elevated temperatures (e.g., greater than 95° C) denature the DNA chains. Reduced temperatures (e.g., less than 63° C) anneal the DNA strands to their appropriate primers. A heat-stable polymerase (isolated from the bacterium *Thermus aquaticus* YT1 which normally thrives in hot springs) allows for an extension of the primers with multiple duplication of the target areas. The newly created double-stranded DNA can be denatured, and the entire process repeated. See Figure 6.2 for a diagram of the "first cycle."

The amplification process is repeated with the unzipping, zipping, and primer extension for a set time period (e.g., 130 minutes for *Legionella* samples), allowing enough time for 30 to 40 cycles of amplification. After the amplification process, millions of copies of a specific DNA sequence are synthesized from each original target. This is generally in excess of 3.4×10^9 copies for each original target DNA. By now, the uniquely identifiable target DNA blueprint has been greatly enhanced and amplified. Detection thus become manageable![6]

The final step is detection. This involves placing the amplified product into special wells with detection strips which have probes (i.e., oligonucleotides that have been affixed to a substrate strip) attached to a predesignated surface site of each strip. In the case of the EnviroAmp™ Legionella Kits, there is a site for the genus *Legionella*, the species *Legionella pneumophila*, and two control sites (e.g., *Legionella* positive and negative representations).

The amplified biotinylated-PCR products attach to the probes, and the wells are washed to remove whatever has not been recognized by the probes. The washed wells are then treated with an enzyme conjugate (e.g., horseradish peroxidase-streptavidin or HRP-SA) and washed again. After a final washing, the substrate (e.g., tetramethyl benzidine) is added to each well. In the presence of enzyme conjugate, a blue color develops. The intensity of this color is compared with the positive control which represents a concentration of 1000 organisms/ml. A slightly noticeable change represents low levels, and a deep blue color change represents high levels. See Figures 6.3 and 6.4.

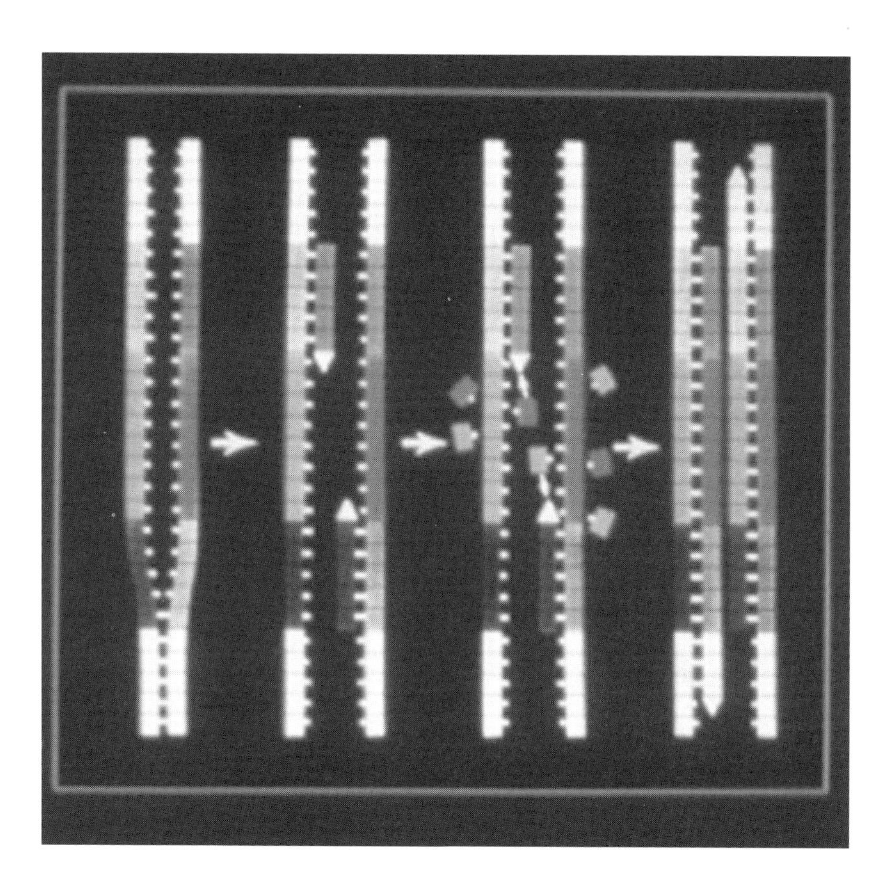

Figure 6.2 Schematic of the "first cycle" of the PCR Process. (Courtesy of Roche Molecular Systems, Inc., Branch-burg, NJ)

The entire process takes around six hours to perform, a fraction of the time required for culturing. It is highly specific, extremely sensitive, and semiquantitative.

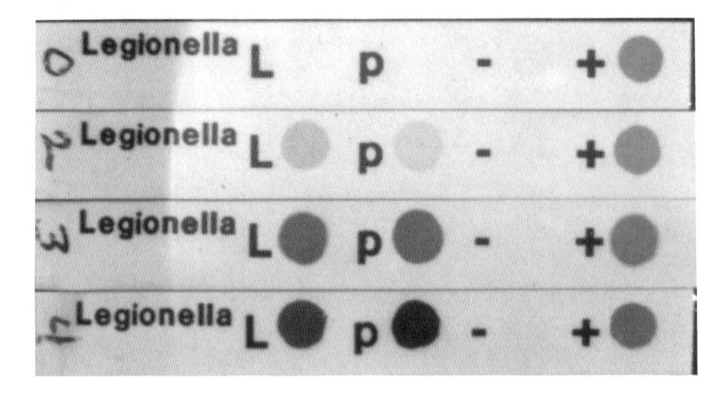

Figure 6.3 Schematic depicting results of amplification of DNA from *Legionella* and non*Legionella* species with (top to bottom) zero organisms/ml., less than 10^3 organisms/ml., equal to 10^3 organisms/ml., and greater than 10^3 organisms/ml. (Courtesy of Roche Molecular Systems, Inc., Branchburg, NJ)

Specificity

The EnviroAmp™ Legionella Test Kit has been tested for false positives against numerous species of bacteria and found to be slightly positive in extremely high amounts of *Altermonas rubra* which is found only in salt water marine sources. The process does detect:

- Twenty-five different species of *Legionella*
- Serogroups 1 through 15 of *Legionella pneumophila*

Quantitation of three species is poor (e.g., *Legionella spiritensis, quinlivanii,* and *sainthelensi)*, and one results in a false negative (e.g., *Legionella israelensis*).

Sensitivity

Sensitivity is on the order of 10 to 100 organisms per ml. of an original water sample. The sensitivity can, however, be improved and brought down to 1 to 10 organisms per ml. by filtering the original sample is down from 100

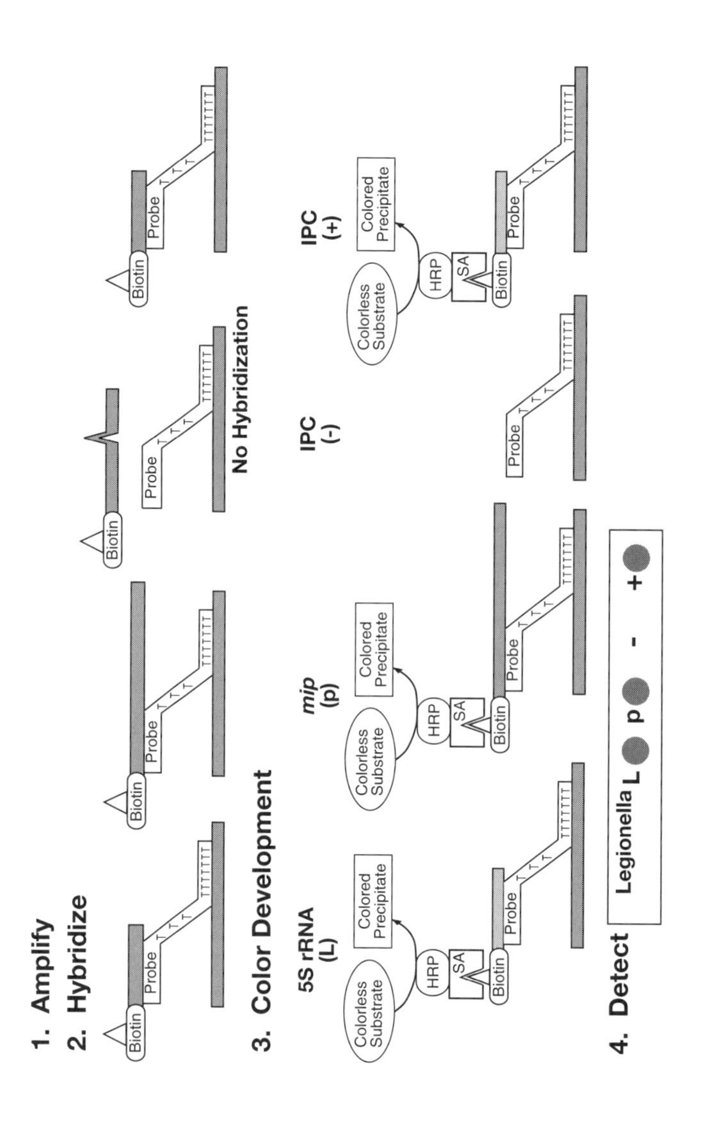

Figure 6.4 PCR amplification and detection of DNA from organisms of *Legionella* using the reverse dot blot detection system. (Courtesy of Roche Molecular Systems, Inc., Branchburg, NJ)

ml instead of the usual 10 milliters. The quantitative results are based on a 10 milliliter sample. So, the results of a 100 milliliter water sample must be multiplied by ten. Increasing the volume of the water sample may also result in increased concentrations of inhibitors to the reaction. This is especially true where the water samples contain noticeable amounts of particulate matter.

Semiquantitative

Quantitation is based on intensity of color change. The break points are as follows:

- Less than 10 organisms/ml. (equivalent to no detection).
- Between 10 and 1,000 organisms/ml.
- Greater than 1,000 organism/ml.

Occasionally, the strips are declared "noninterpretable." In these cases, the controls either failed to respond as anticipated or responded when they should not have. This may be due to the presence of amplification inhibitors or improper processing. In some cases, these problems can be rectified (e.g., diluting a sample or redoing a step in the processing). Where there are no responses to any of the four sites, there may be an excess of inhibitor. Additional steps may be taken at an additional expense to further remove inhibitors from the sample(s).

Commercial Laboratories

There are presently around twelve environmental laboratories capable of performing the analysis of *Legionella* by PCR. These may be identified through the distributer of the equipment, i.e. The Perkin-Elmer Corporation in Alameda, California. The Food Environmental Group Leader in Marketing Administration maintains a listing of laboratories.

Interpretation of Results

Controversy surrounds the interpretation of PCR results for use in the environmental field. Identification of a specific microbe is just that. Its presence has been confirmed, and semiquantitation is of viables and nonviables. Whereas some feel the inclusion of nonviables can be misleading, others have used the levels successfully in responding to outbreaks of disease and ascertaining which sources have elevated levels. Furthermore, there has been speculation that Pontiac fever may be caused by elevated levels of nonviables as well as those which are invasive. Where confirmation of numbers and viability is desired, backup culture sampling may be performed while sources can be identified and corrected within a relatively abbreviated time period.

A rapid means for locating contaminant sources, the PCR results are not recommended for routine, nonemergency use in evaluating office and workplace environments for elevated levels of pathogens.[7] This may change with further research and as the methodology for *Mycobacterium tuberculosis* is developed and refined. Yet, the results for now are primarily for detecting low, moderate, and high levels of *Legionella spp.* and *Legionella pneumophila* in water reservoirs.

BACTERIAL IDENTIFICATION[8]

Bacteria are easily identified by the Microbial Identification System (MIS) which is also referred to as the Fatty Acid Methyl Esters (FAME) Process. It is a fully automated gas chromatographic analytical system which identifies bacteria based on their unique fatty acid profiles. The time required for analysis is primarily that of culturing the sample, extracting a suspect colony, and processing it through a gas chromatograph which has been coupled with a database which compares the analytical information with over 2,000 species.

Upon completion of the initial culturing process, laboratory technicians typically must go through a long, tedious process of staining the colony (or colonies), observing them microscopically, and culturing the individual colonies on different media again. The MIS, however, circumvents this requirement.

Not only is one colony more rapidly, easily identified down to strain with minimal costs, but multiple samples can be run simultaneously with relative ease. Also the cost for performing MIS is a fraction of that required for the old method. The MIS is objective and reproducible whereas the old method required an experienced microbiologist to spend hours and days culturing the microbes. With MIS, microbiological identification has been simplified, speeded up, and less costly.

Uses for the MIS

The more obvious use for the MIS is easy species and strain typing of *Mycobacterium tuberculosis* and *Legionella pneumophila*. The bacterial strains which are associated with antibiotic-resistant *Mycobacterium tuberculosis* and different forms of the genus *Legionella* are not only differentiated but by species but by strain as well. Although is takes several days to culture and subculture these pathogens, strain identity can also be used to track sources. For instance, a strain or strains of a given species of

Legionella which has been isolated from a cooling tower may be different from those in a humidifier.

Illness caused by various strains of *Mycobacterium tuberculosis* may also be tracked back to its origin. In other words, an outbreak of tuberculosis may be tracked clinically to the source, and control measures may be instituted at the source instead of through generalized, all-out response actions at all potential locations. The latter can be an overwhelming task where a single identified source is more manageable.

The pathogenic microbes pose a life threatening problem for which solutions may involve life and death situations. Yet, there are uses for non-pathogenic bacterial fingerprinting as well. These are generally industry related.

In pharmaceuticals, supplies and equipment requiring sterilization occasionally are contaminated with bacteria resistant to the treatment. After taking a surface sample and culturing, bacterial growths can be identified and corrective actions taken to control odd strains may be initiated. An example of where this is frequently used is in pig heart sterilization following gamma radiation treatment. Surface samples disclose microbes which are resistant to radiation. Bacteria are ubiquitous and varieties of different bacterial strains are resistant to heat and/or ethylene oxide treatment.

In the semiconductor industry, particulate contamination in a clean room may be due to bacteria laden skin cells created by employees (primarily those with dry skin) or contaminated surfaces where microbes may be residing. Each individual has his or her own mix of bacterial strains which may differ from other individuals, and the strains are body-area dependent (e.g., *Staphlococcus* strains found under the arm pits differ from that in one's scalp). Thus, the individual or a surface responsible for contaminating a clean room may be tracked. When individuals are suspect, their outer garments may submit for testing upon departure from the clean room. When a particular individual or individuals are identified as potential sources, sampling is often narrowed to various parts of the body.

In outdoor environments, air, soil, and water contamination may require strain identification as well. Interestingly, strains are the same worldwide. They vary, however, in their fatty acid make-up in the nutrients upon which they grow, and microbes can grow under some incredible conditions. Some microbes grow well in formaldehyde and phenol. Gold mining operations utilize cyanide extraction, yet there are some strains of bacteria which consume and decompose the cyanide. Where the mining operation may trust the process, extractable gold may go uncaptured due to the presence of a given microbe. Some microbes grow on stainless steel substrates, and their metabolites damage the alloy. Some grow in the presence of low levels of mercury and are capable of methylating the mercury. Product and building materials trouble shooting may be performed through strain identification. Strain tracking can also be a useful tool in environmental contamination by industry.

Sampling Methodology

Sampling will typically be performed indoors. Occupational exposures will require mostly air sampling, and environmental sampling will require surface sampling.

Both types of sampling are performed in the same fashion as directed in Chapter 3. Ultimately, the samples are collected onto or plated onto a culture media retained within a Petri dish. The sought after microbe must be known and the culture media may or may not be the same as that which has been previously indicated. The laboratory which is to perform the analysis must be consulted during the planning process. Even if the proper culture media is indicated within this chapter, the environmental professional will need to know the type of database used by the laboratory, and the laboratory may have a special adaptation to the methods set forth by normal protocols. The sampling strategy and methods remain the same.

With the exception of culture media, proceed to sample as previously directed. The difference is in the analysis.

Analytical Methodology

More than 300 fatty acids and related compounds have been found and are in different combinations or absent based upon strain. The other compounds include aldehydes, alkanes, and dimethyl esters. All are lumped into one group and are, for the sake of simplicity, referred to as "fatty acid methyl esters."

The process involves culturing, sample processing, and gas chromatograph identification. See Figure 6.5 for an overview of the extract-processing methodology and Figure 6.6 for the analytical methodology.

Culturing

The fatty acid composition of microbes is affected by temperature and growth media. For this reason, the suspect microbe(s) must be cultured according to their classification. Some of the different classes and culturing requirements are as follows:

Aerobic bacteria

- TSBA Database—Environmental broad database which includes environmental and clinical aerobes.

Environmental aerobes: Trypticase soy broth agar at 28° C

- *Legionella* Specific Aerobes Database—Buffered charcoal yeast extract.

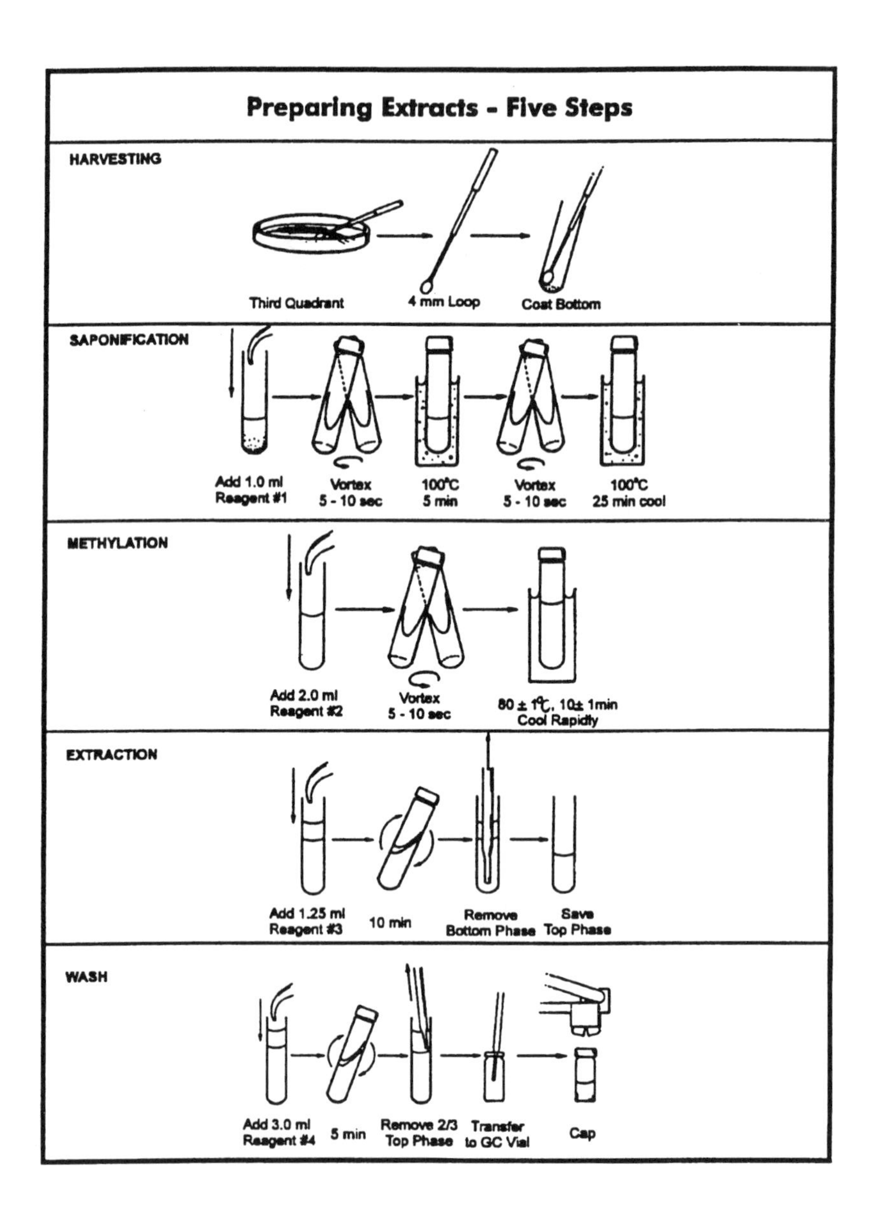

Figure 6.5 Five step extract-processing for use in the Microbial Identification System. (Courtesy of MIDI, Newark, DE)

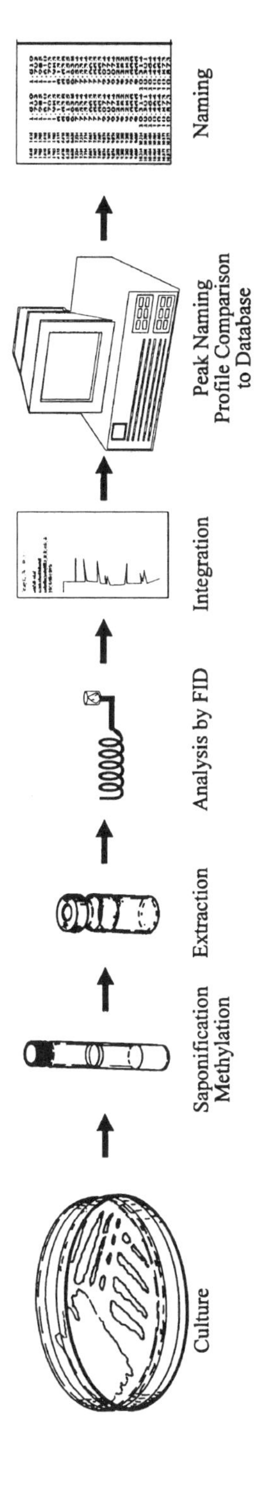

Figure 6.6 Overview of the Microbial Identification System analytical methodology. (Courtesy of MIDI, Newark, DE)

- CLIN Database—Clinical database which includes only the clinical aerobes, typically requires blood agar with a trypticase soy broth base at 35° C.

Anaerobic bacteria

- Anaerobic Database—Brain-heart infusion with supplements at 35° C.
- VPI Anaerobic Database—Peptone yeast glucose broth at 35° C.

The amount of time required for growth is typically established so as to reduce the effect of age differences. The microbes are grown initially on any of a number of culture media but later subcultured (transferred to another culture media) which is specific for the bacterial classification for which there is an expressed concern.

The specificity requirements of the nutrient is exacting, down to the brand of nutrient media purchased and to the moisture content. Then, too, some bacteria which have been previously grown on a different nutrient may require two or three generations of subculturing to obtain consistency of composition. Consequently, the colony age and nutrient constituents are tightly controlled.

Culture times generally take twenty-four hours for aerobes, forty-eight hours for anaerobic bacteria. There are, however, a few exceptions to the norm. One is *Legionella.*

Legionella takes two to three days for a sufficient amount to grow in order to replate and/or analyze the sample. Keep in mind that should a second culture may be required, and the amount of culturing time may double. Where *Legionella* is in question, both quantitatively and qualitatively, the original sample may be cultured for two or more weeks. Then another subculture sample must be taken from the new colony and plated onto different media for strain identification. Thus, the culturing time required for *Legionella* is a major drawback to using the MIS in this particular instance.

Sample Processing

After grown for a specified time, on a specified nutrient, a bacterial colony is harvested (or taken) from the plate and placed into a test tube. In the test tube, each sample undergoes saponification, methylation, extraction, and washing. Saponification cleaves the fatty acids from their lipids. Methylation adds methyl alcohol to the fatty acids to convert them into fatty acid methyl esters. The water insoluble fatty acid methyl esters are then extracted with hexane, and the remaining components are cleaned from (or taken out of) the sample.

Gas Chromatographic Identification

The gas chromatograph is specially fitted with a Microbial Identification System software package. The software library is capable of recognizing over 2,000 strains of bacteria by their fatty acid components and relative amounts. Some adjustment is made within the system to make allowances for differences which may result from slight temperature and age variations.

The data banks have been compiled from cultures which have been collected from around the world to avoid that which was potentially geographic biases. These biases were later found not to exist.

The "comparison standard" is a mixture of straight chained, saturated fatty acids from nine to twenty carbons in length with five hydroxy acids. Known amounts of the standard mixture are evaluated each time an analysis is run. The standard consists of only 17 of the 300 possible fatty acids. This seems to be adequate for comparative purposes with a majority of the samples.

The MIDI Computer Database is continuously undergoing updates and add-ons to the 2,000 stains in the MIS. Yet, the database still falls short of identifying "all bacteria." Some may be listed as "unknowns" or unconfirmed attempts at fingerprinting the genus, species, and/or strain. The well-initiated laboratory analyst may be fairly accurate at these unconfirmed attempts. Attempts are typically guilty-by-association.

The analytical data retrieved from the chromatograph provides the standard name, amount of each FAME, and the strain of greatest fit. One limitation is where a sample has more than one strain. Efforts are underway to circumvent this problem and fit the "top three suspects." An example chromatographic printout is presented in Figure 6.6.

Commercial Laboratories

There are limited commercial laboratories available to perform this process, and they are predominately clinical labs. Those laboratories which have the facilities to perform the work may be identified through those who sell the database and calibration standards. These are MIDI which is located in Newark, Delaware and Hewlett Packard.

Interpretation of Results

The results interpretation is similar to that of Chapters 3 and 4. The only difference is greater specificity and clarity not only as to the genus and species but the strain of bacteria as well. Fingerprinting becomes more exact, less time consuming, less subjective, and less expensive than culturing and special laboratory sample manipulations.

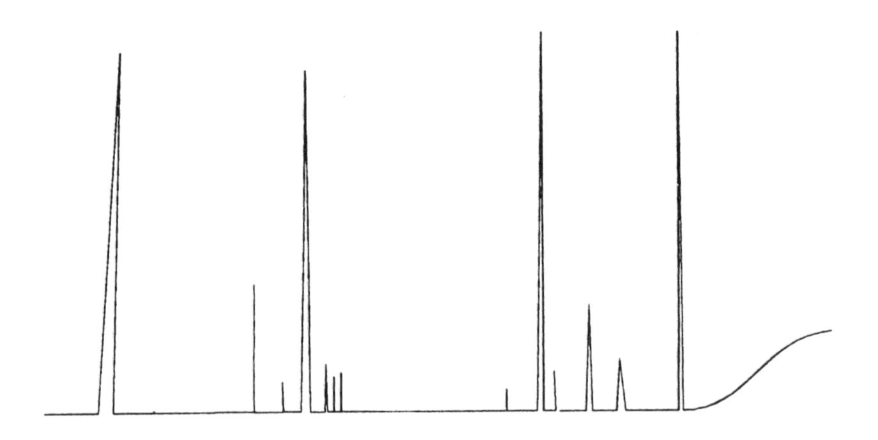

	NAME	AREA %
1.	C14:0 ISO	4.52
2.	C14:0	1.10
3.	C15:0 ISO	9.96
4.	C15:0 ANTEISO	37.97
5.	C16:0 ISO	2.53
6.	C16:0	5.06
7.	C17:0 ISO	3.33
8.	C17:0 ANTEISO	6.02
9.	C18:0 ISO	0.71
10.	C18:0	14.48
11.	C19:0 ISO	0.70
12.	C19:0 ANTIESO	0.73
13.	C20:0	11.89

Figure 6.7 Sample gas chromatographic run and partial computer printout of MIDI Microbial Identification System (Courtesy of MIDI, Newark, DE)

REFERENCES

1 Perkin-Elmer. EnvironAmp™ Legionella Kits. [Procedure booklet] Roche Molecular Systems, Inc., Alameda, California, 1995.
2 Application of Polymerase Reaction to Environmental Monitoring. Bulletin. Reprint from *Envrionmental Laboratory*, December 1993.
3 Mullis, Kary B. The Unusual Origin of the Polymerase Chain Reaction. *Scientific American*, New York, NY, April 1990.
4 Thio, Claudia. "Uses and Understanding the PCR Test Kits." [Oral communication] Environmental Products Division, The Perkin-Elmer Corporation in Almeda, California, August 1995.
5 Schafer, Millie P., Ph.D. Sampling and Analytical Method Development for Airborne *Mycobacterium tuberculosis*. [Proposal] NIOSH, Cincinnati, Ohio, Revised August 30, 1994.
6 Thio, Claudia. Application of Polymerase Chain Reaction to Environmental Monitoring. Reprint from American Environmental Laboratory, December 1993.
7 Morris, George. The use of PCR methods verses culturing. [Oral communication] PathCon Laboratory, Norcross, Georgia.
8 Sasser, Myron. Identification of Bacteria by Gas Chromatography of Cellular Fatty Acids. [Technical Note #101] MIDI, Newark, Delaware, May 1990.

Chapter 7

FORENSICS OF ENVIRONMENTAL DUST

Although microscopy seems by many to be a technological dinosaur, it is limited only in its failure to be recognized for its versatility. It is of interest that Dr. Walter C. McCrone dated the Shroud of Turin in 1980, based solely on observations made with a polarized light microscope which he confirmed later by X-ray diffraction and electron microprobe analyses. Pigment and cloth samples had been lifted by tape from the Shroud in order to leave it fully intact and undamaged. Dr. McCrone concluded that the image had been painted just prior the Shroud's first exhibition in 1356. Yet, his conclusions left him a very unpopular man! He was scorned as a charlatan by those who wished to believe in the Shroud's authenticity and became the recipient of hate mail. Later, in 1988, the Universities of Arizona, Oxford, and Zurich Technology Institute studied the Shroud using "carbon dating." The University dated the Shroud around 1325 $\pm$ 65 years (not at the time of the crucifixion of Jesus) and within the time period pinpointed by Dr. Walter C. McCrone—using only a microscope![1,2]

Since as early as the late 1800s, microscopy has been used by the forensic scientists for crime detection. Pollen typing has been used recently by some forensic scientists in determining the source areas for illegal shipments of marijuana. Soil samples are used to determine the source area of crime scene contaminants. Clothing fibers are traced by the type of fiber and special dyes. Hair can be differentiated as to species (e.g., human, dog, or cat) and distinct color, texture, and thickness. Dust on articles found at a crime scene may contain evidence of an association with certain industrial operations which are distinct and recognizable.

All the materials addressed in forensic microscopy have environmental and occupational exposure correlations. They are not separate and distinct. Where environmental unknowns are an issue, an initial evaluation of the dust is a direct, nondestruct approach to obtaining necessary information.

Components of air, settled dust, and water samples may be characterized rapidly without the need for the guessing-game which is generally required when substances are chemically analyzed. Like differentiating a house from a shed or a dog from a cat, microscopic particles are immediately classified by an experienced microscopist into one of several categories, then further isolated and identified by various microanalytical methods. Figure 7.1 is a generalized photomicrograph of dust with a representation of many of the categories which are mentioned in Table 7.1 and some representative examples.

Once categorized, microanalytical methods used for identification more often than not may be performed by nondestructive examination of their morphology with as great an accuracy as and certainly less sample size than that of chemical analyses, and the chemical structure of a given particle or particles may be determined as well using a little more complicated methods. All particles can be identified by one of or a combination of these other supplementary microanalytical methodologies, and sometimes microchemistry is used on material which would not ordinarily be identified due to the limited size and/or volume of material available to analyze. Consequently, forensic microscopy is finding its way into the environmental professional's tool box of options.

ENVIRONMENTAL USES FOR FORENSIC MICROSCOPY

Forensic microscopy has been inching its way into importance within the environmental air quality and indoor air quality arenas only in the last ten to twenty years. Where other tools have been limited, microscopy has filled the gap. Yet, its use has still been relegated mostly to research. The environmental professionals have yet to explore the full potential of this very powerful tool. Those few who have chosen this approach are discovering new uses every day. A few of these experiences are described herein.

Table 7.1 Microscopic Characterization of Environmental Dust Components and Some Distinct, Identifiable Materials

BIOLOGICAL
- Pollen
- Fungal and bacterial spores
- Algae
- Insect parts
- Skin cells

FIBERS
- Hair (e.g., human, cat, or dog hair)
- Clothes fibers
- Paper fibers
- Spun fibers (e.g., glass fibers)
- Mineral fibers (e.g., asbestos)
- Wood (hard wood versus soft wood)
- Plant fibers (e.g., seed hairs, blast/leaf/grass fibers)
- Miscellaneous (e.g., carbon fibers, feathers, spider webs, etc.)

Table 7.1 (continued)

MINERALS
Soils
Amorphous versus crystalline
ENVIRONMENTAL POLLUTANTS AND INDUSTRIAL PRODUCTS
Soot and ash
Metal fumes
Paint
Explosives
Drugs

Excerpted from *Forensic Microscopy.*[3]

Environmental Pollution

It has been stated that "the world's great ice sheets and glaciers contain memories (in their time-deposited layers of ice, dust, and debris) which trace the effect of human activities on our atmosphere."[4] Likewise, interplanetary dust and comet material record the primordial processes that occurred in the outer regions of the solar system long before the earth was formed. On a much smaller scale, it is understandable that dust analysis can provide information concerning the environment over shorter time periods, imaging the present as well as the recent past. Despite the phenomenal capacity of forensic microscopy, environmental information and data collection is rarely accomplished by deposition and characterization of dust.

Analyses of environmental pollutants were first performed by Dr. Walter C. McCrone in the late 1960s. As he used forensic microscopy quite successfully in recreating a crime scene, Dr. McCrone envisioned an even greater scale. Identification of atmospheric particles has since struck a cord of intrigue, and some of the astonishing results serve as a testament to the efficacy of such a practice.

Forensic microscopy has been used to address industrial neighborhood complaints. In one instance, a resident in the vicinity of industrial activities complained of a sticky substance collecting on his automobile. He speculated that the source was a nearby steel mill. Then, the neighbors began to complain. The steel mill management responded swiftly, stating that "the natives are restless." They did not think the source of the sticky material was their plant. A folded sheet of paper with the implicated substance was submitted to a laboratory with a rush request. The microscopist had the answer in minutes. He called the steel mill which requested the analysis and asked about the types of trees growing in the area. Their response was, "What does this have to do with the problem?" "Simple," responded the microscopist. "These trees are pollinating at the moment. The samples submitted consist(ed) entirely of

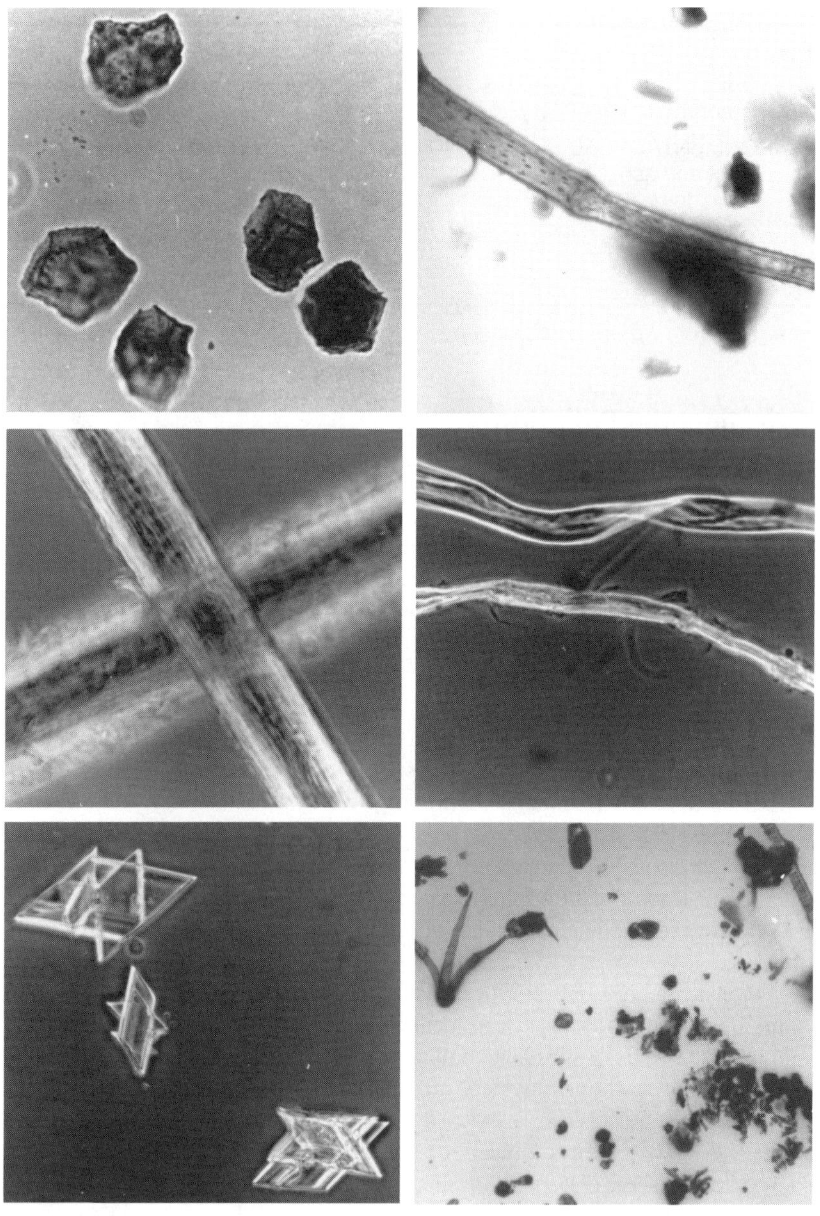

Figure 7.1 Photomicrographs of environmental materials, magnified 400x. They are epidermal cells (top left), an insect leg (top right), hair fibers (middle left), clothing fibers (middle right), crystalline mineral formations (bottom left), and a general overview of environmental dust (bottom right). The latter shows minerals, spores, wood fiber, pollen, and plant hair.

pollen from the (trees found in the area)." A quick resolution was forthcoming.[5]

In air emissions monitoring, particulate material is sampled within the exhaust stack and at various sites remote to the stack. Although emissions generally require chemical analysis, particles are weighed, sometimes analyzed by microscopy for particulate size distribution. Where additional information is sought by the environmental professional, the particles may be analyzed for composition.

Composition of hospital incinerator emissions may be evaluated microscopically for burn efficiency of biological materials. Foundries' exhausts may be evaluated for fumes and minerals. Fiber glassing operations may be checked for fiber discharges. Waste management incinerators can be evaluated for all materials (e.g., biologicals, fumes, minerals, and fibers). Prior to installation, control equipment is frequently chosen on the basis of emissions testing without any controls in place. Later, the efficiency of the equipment following installation is ascertained to evaluate its efficacy.

Microscopic examination of the material before and after is compared for particulate amount, size, and type distribution. By this method, not only can the level of efficiency be determined, but the reason for inefficiency may be ascertained as well. Certain particle sizes may be escaping capture. An incinerator may not be burning hot enough to reduce the emissions to gaseous form nor to completely incinerate certain materials. There may not be enough oxygen in the burning process, or other components which were not anticipated are being discharged. The diagnostic competence of microscopy is only beginning to find its way into environmental pollution evaluations.

Microscopic techniques have been used for before and after sampling and analysis of industrial emission controls. In one case, a large industrial incinerator fused submicron-size particles (e.g., smaller than 1 micron) to form larger ones (e.g., 10 to 30 microns). The elemental composition (as determined by electron microprobe) showed a high percentage of silicon and other elements (e.g., titanium, iron, sulfur, and aluminum with some calcium, chlorine, and potassium). The fused particles were identified as fused amorphous silica with traces of other the elements. Thus, the product created by incineration was fused amorphous silica, greater than 10 microns in diameter, above the respirable particulate range.[6]

In one situation involving the attorney general of Illinois, a consultant was hired to collect and analyze suspended and settled dust samples outside a given lot line. The firm collected leaves which had settled dust in the vicinity of a ferrous metal foundry. Then a three week accumulation was collected from the surface of undisturbed snow within the same vicinity as where the leaf samples were taken. The samples collected from the leaves showed the settled dust to be composed mostly of particles of iron oxides (magnetite and hematite), calcite (limestone), quartz, and clay (bentonite) which are typical emissions from a foundry. Lesser amounts of fuel (bituminous coal dust) and combustion products (coal boiler flyash and oil soot) were also found. Then, the three week accumulation snow sample was found to be of similar material. The foundry was thus implicated as the source of environmental dust accumulation. As 98 percent of the dust was determined to have originated from the foundry, air sample results could likewise be correlated to the amount of dust contribution from the plant. The information was sufficient to gain access under warrant.[7]

Table 7.2 Typical Analysis of Urban Air

Substance	Source	Percentage (< 2 µm in Size)	Percentage (> 2 µm in Size)
quartz	soil sand	0	20
quartz	foundry sand	0	5
limestone	soil	2	8
feldspars	soil	0	5
iron oxides	rust	2	2
iron oxides	soil	2	0
clays	soil	7	0
oil soot	combustion	1	5
coal (unburned or coked)	combustion	2	4
glassy flyash	oil burner	3	0
glassy flyash	power plant (coal)	0	7
incinerator	combustion	0	5
pollen/spores	plant life	0	2
plant fibers	plant cutting	0	2
paper fibers	paper scrap	0	3
rubber	tires	0	8
magnetite	power plant	0	2
magnetite	soil	--	1
paint spray	industrial	--	--

Excerpted from *Industrial Research.*[8]

Sediment in rivers and harbors may be fingerprinted by composition. The identity of the material and its sample sites can assist the environmental professional track the source of industrial effluent into the waterways. South Lake Michigan is characterized by steel mill effluent. New Jersey rivers are distinct in their zinc-containing minerals. The denial of a generator that a given effluent belongs to a neighboring plant of similar production processes can be circumvented with the simple use of microscopy.[9] Although their effluent may the similar, it is "not the same."

Analyses have been performed to determine the source of lead dust where elevated lead levels have been determined by instrumental analysis (e.g., atomic absorption spectrophotometer). Instrumentation can accurately determine levels but not the source. Some of the potential sources include, but are not limited to, lead-based paint, leaded-gasoline exhaust contamination, flyash, and natural lead soil deposits. Where lead-based paint has been identified as a source of elevated exposures, the paint can further be analyzed to determine which layers contain the lead, year(s) produced, and possibly the manufacturer. The latter information may be useful in litigation.

The uses of forensic microscopy for particulate analysis in the environmental field are limited only by oversight. It is up to the environmental professional to explore the capabilities of this approach and extract that which may prove helpful in a given situation. Table 7.3 contains an abbreviated listing of substances classified as industrial dusts and combustion products which can be identified, sized, and quantitated by microscopy.

Table 7.3 Environmental Pollutants Identifiable by Forensic Microscopy

INDUSTRIAL DUSTS	
	Abrasives and polishes (e.g., silicon carbide)
	Catalysts (e.g., large particulate metals)
	Cements (e.g., Portland cement)
	Detergents (e.g., organic sulfates)
	Fertilizers (e.g., nitrogen, phosphorous, and potassium)
	Food processing (e.g., caffeine in coffee)
	Metals refining and processing (e.g., copper)
	Pigments (e.g., cinnabar)
	Polymers (e.g., urea-formaldehyde resin)
COMBUSTION PRODUCTS	
	Auto exhaust
	Coal flyash
	Hazardous waste incinerators
	Oil soot
	Smelters
	Trash incinerators

Excerpted from Forensic Microscopy.[10]

Indoor Air Quality

One of the first documented indoor air quality analyses which was performed by microscopy was accomplished by The McCrone Institute in Chicago, Illinois. This was to determine the potential for contamination of preparation slide by settling dust in one of their laboratory areas. They found that "nearly 1,000 particles," larger than 5 microns, were settling every hour on each of several bench sample sites of one square centimeter (e.g., 24,000 particles per square centimeter per day). The particles, which represent typical human habitats, were identified as human epidermal cells, plant pollen, human/animal hairs, textile fibers, paper fibers, minerals (from outdoor soils and dust brought indoors), flyash (from the gas-burning furnace in the building), and a host of other materials used/found within their laboratory environment.[11] This approach can likewise be used to categorize particulate materials found in "tight building syndrome" cases.

A recent study by Cornell University suggests that tight building syndrome may be caused by glass fibers.[12] Although it has become controversial, the issue of fiberglass fibers being implicated in indoor air could only have become a viable candidate for consideration through microscopic analyses of environmental dust. Possible sources of airborne glass fiber exposures include, but are not limited to, fireproofing in air plenums, ceiling tiles, duct board, and furnace filter material. See Figure 7.2 for some different examples. Again, others are implicating pollens, spores, and fungi. All require microscopy for initial identification and sometimes quantification. These have been discussed in Chapter 3.

Microscopic analyses of indoor particles have been addressed in residences. In one instance, a woman complained of a home-related itch which her doctor speculated was probably caused by glass fibers. McCrone's Laboratory was asked to perform an evaluation. Settled dust samples were collected from various areas around the house as well as building/furnishing materials known to have fibers. Each of the settled dust samples was found to contain large amounts of glass fibers which were impregnated and covered with globules of a pink resin. None of the known material samples appeared to have the same appearance. They returned for additional samples and located the residential insulation (e.g., batting) in the enclosed wall spaces. This material was confirmed to be the culprit. Upon contractor removal of the material, the means by which the insulation was entering into the residential air stream was identified. There had been a "flaw" in the air handling ducts where the air movement was able to collect and distribute the associated insulation glass fibers. The problem had been tracked by microscopy.[13]

Microscopy provides a means for identification of particle sources in the semiconductor cleanrooms. Particle counters may indicate an excessive level of particulates, but only microscopy can identify the source. Paper fibers may indicate that notebooks, stock records, inventory cards, and other sources of paper should be eliminated from the room or kept in plastic envelopes. Dyes and other distinctive additives to the paper can further aid in isolating the source.

Dust sampling in buildings known to have asbestos-containing building materials has also been performed. In 1935, rafter dust samples were taken in

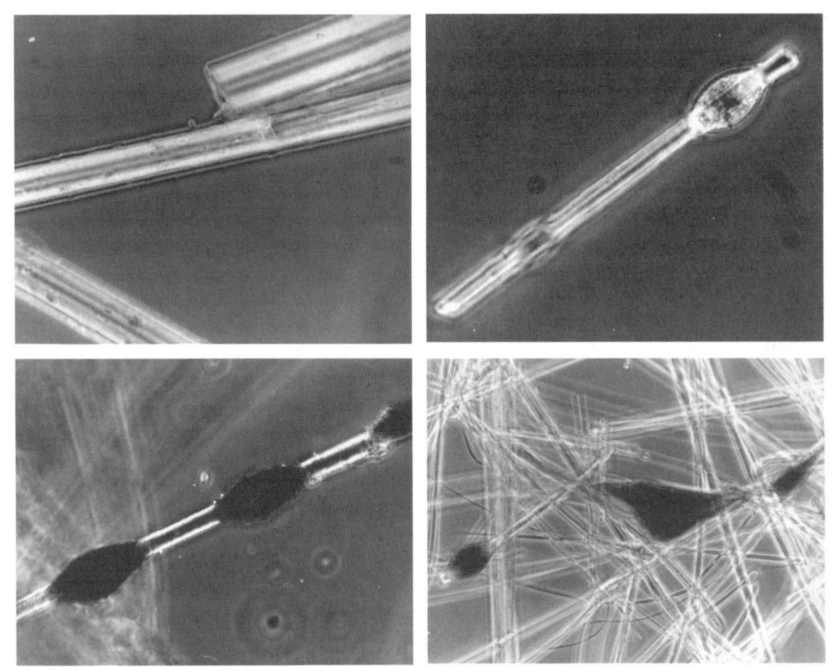

Figure 7.2 Photomicrographs of fiberglass fibers from different sources, magnified 400x. They include: (top left) untreated fiberglass, (top right) duct board, (bottom left) duct board with coating material treated using xylene and sulfuric acid to affect a color change that tags free aldehydes, and (bottom right) thermal insultation with asphalt-impregnated binder.

various asbestos mills for characterization of fibers by phase light microscopy. In 1983, it was reasoned that:

> "Clearly, loose dust of any nature can become resuspended in air, and there must be a level of surface contamination at which the concentration of resuspended fibers, in the particular case of asbestos, approaches the recommended limits for breathing air."

In 1986, the National Institute for Occupational Safety and Health (NIOSH), along with the West Virginia Department of Health, presented a paper on the use of settled asbestos-contaminated dust. In 1988, McCrone Environmental Services reported values of asbestos in settled dust from 10,000 to a billion structures per square foot as analyzed by transmission electron microscope. Collection and analysis by microscopy of potential contaminated dust have since been receiving wider acceptance and use.

SAMPLING METHODOLOGIES[14,15]

Sampling methodologies are oftentimes dictated by the laboratory which is going to perform the analysis. Although there are a few published approaches, most procedures are worked out between the analytical laboratory and the environmental professional. Some of the methods mentioned herein are the product of the experience of one of the more experience laboratories in environmental analyses. Others are excerpted from various publications

Settled Surface Dust

Settled dust may be collected from smooth surfaces (e.g., desk tops) and rough surfaces (e.g., carpeting) by any number of creative techniques. Some require specialty supplies. Others require the use of that which is readily available at a local retail store.

Specialty Tape

Specialty tape may be purchased from microscope supply venders. The tacky material is minimal and does not hold the collected material such that it becomes difficult or impossible to remove. The taped material is retained by affixing the tacky surface to a clean surface and placed in a fiber-free envelope/baggy for transport to the laboratory. At the laboratory, the microscopist will "pluck" the material from the surface of the tape by using a special micromanipulation device (e.g., fine tungsten needles with a tip measuring 1 to 10 microns in diameter). With this technique, the dust components may be isolated and identified individually.

Transparent Tape

A transparent tape which is available in some office supply stores is easy to obtain. This tape is not the usual tape which one can see through when it is affixed to paper. It is a clear tape where the writing on the carrier is readable. The tape is touched to a surface, and dust particles adhere to the sticky portion of the tape. This is then placed immediately onto a microscope slide with or without a stain.

The drawback to this method is that once affixed to the slide, the collected sample cannot be further manipulated and stained without great difficulty (e.g., treating the surface of the tape with a solvent). Then, too, if there is excessive material on the tape to reasonably distinguish individual particles, the sample may again require special processing. With these limitations in mind, the environmental professional may choose to use this technique only for screening and gross examination of material or for confirming the presence of a suspect material for which the microscope slide has already treated with the appropriate stain.

However, if the environmental professional should chose to use the clear tape, there are means available to manage the otherwise irretrievable sample. The particles which have adhered to the tape may be removed by lifting the tape, applying a small drop of benzene, and using a fine needle to make a small ball of the adhesive trapped particle(s).[16] The trapped material can be withdrawn and the ball of adhesive removed chemically. This process is tedious and choice of a more easily manipulated collection media is desired whenever feasible.

Post-it Paper [15]

Post-it paper is excellent for sample collection as it is easily obtained and inexpensive. The sticky surface of a Post-it is pressed onto the settled dust, the paper folded into itself (with the sticky portion inside), and shipped to the laboratory of choice in a plastic zip-lock bag. Analysis may be performed by particle-picking of material from the sticky surface or by scanning electron microscope while the particles are still on the paper.

Dust Collection Surfaces[18]

A thin film of Aroclor 1260 is placed on a microscope slide and the treated slide placed where dust may collect and adhere to its surface. An alternative is the placement of the sticky side of transparent or specialty tape outward so dust will collect on its surface.

The alternate transparent tape is not only more flexible for handling, but it also provides a backing which may be wrapped around a structure (e.g., a tree) or extended to include a larger sample area (e.g., on top of an office space separator, extending from under the air supply vent to a stagnant air pocket). Reference directions should be clarified on the nonsticky side of the tape.

Collection efficiency is based upon particle size, air velocity (e.g., wind speed), and direction of impact (e.g., air blowing at a 45 degree angle to the surface). These influencing factors should be recorded, along with the reference information, on a separate field notebook.

Microvacuuming[19]

Microvacuuming has been receiving a considerable amount of attention, particularly for asbestos-contaminated settled dust. A vacuum pump is used to collect dust particulate within a 100 square centimeter surface at a recommended flow rate of 2.0 liters per minute. This method is particularly useful in dust collection from irregular surfaces (e.g., carpeting).

The detection limit of this methodology is 150,000 structures per square foot [or 161 structures per square centimeter (structures/sq. cm.)] as determined by transmission electron microscopy (TEM). Concentrations over 1,000 structures/sq. cm. are considered elevated, while levels over 100,000 were used to indicate an abatement project barrier has been breached. There are no regulations which provide acceptable/nonacceptable limits.

Airborne Dust Sampling

Airborne dust capture may be preferred over settled dust collection in order to determine the existing suspended particulates (not the existing and previously deposited) and to collect some of the smaller material which may not have settled out from or become resuspended in the air due to the size and/or shape of the material of possible concern.

Membrane Filters

Air sampling may be performed for suspended dust by using a membrane filter and air sampling pump. Recommended filter types include, but are not limited to, the following:

Polycarbonate filter—Using a stereo microscope, the microscopist may selectively isolate and pluck material from the surface of the filter.

Fiberglass filter—The microscopist may slice the filter and look at the material through an unspecialized microscope (e.g., not a phase contrast microscope) which will allow a view of the material on the surface of the fiberglass while the light passes through the thinned-out fibrous backing.

Mixed-cellulose ester filter—The microscopist may melt the filter (in a procedure similar to that of asbestos air sample filter analysis) using vaporized acetone. This method is not recommended in most cases where desiccation and/or destruction of the sought after material may occur. If the material is an unknown, this approach will disallow identification of content.

Keep in mind that each of the above filters have varying pore sizes, dependent upon the manufacture and the various specifications. One filter type may come with several choices as to pore size and/or particle retention capabilities. The smaller the pore size, the more expensive the filter. Be certain to obtain one which will capture particulates down to 1 micron in diameter or better. All of the above-mentioned filters can be purchased with a minimum of 1 micron and down to better than 0.025 microns.[19] Unless electron microscopy is to be performed, the latter is unnecessary.

The air sampling flow rate should be adjusted to produce as large a sample volume as possible within the time period desired while keeping the upper limit within a flow which will not cause damage to the filter or desiccate biologicals. In most cases, a flow rate of one liter per minute may be used for biologicals, and fifteen liters per minute should not destroy the filter.

The larger the air volume, the more representative the sample. Although total sampled air volumes have been as low as 100 liters, a collection of 2,000 liters may prove more valuable. Then, too, the biologicals stand a greater chance of being damaged with a longer sample duration. So, the environmental professional may choose to take a minimum of two samples per site where biologicals are a possibility. The two samples may represent the lower range and the larger volume of sample. Based on the method of analysis, in most cases, the air volume cannot be excessive. Like a dump truck of soil sample, the limitation of the air sample volume is relevant only in the provision of a manageable sample size.

Even if the microscopist is not intending to quantify the results in particle per cubic meter, air volumes should be recorded in the off-chance that the volume may be used later (e.g., where asbestos fibers are identified, the airborne fiber counts may be determined). The air volume, however, will rarely be relevant to the microscopist.

Cascade Impactors

An adhesive film may be placed on the surface of each stage of an impactor, and the sample collection time may be limited while separating the collected material by size. As the particles tend to impact singularly, the microscopist may analyze each adhesive film directly. Separation and isolation are already completed by this sampling method.

Others

Other air sampling techniques which have been used include impingers and cyclones. Processing these samples may be more involved (e.g., time consuming and expensive), yet impinger and cyclone samples allow for dilution and separation of the collected material. They are, however, limited in their ability to collect only certain particle sizes. The impinger collects particles greater than 1 micron in diameter, those which are visible under the light microscope. On the other hand, the cyclone collects nonfibrous particles of less than 5 microns. Anything greater than 5 microns and/or fibrous in nature is likely to be overlooked.

Bulk Sampling

An alternative to the specialty tape and problems associated with the clear tape is "bulk dust" sampling. Where dealing with large deposits of dust (or dirt), bulk sampling becomes the most feasible approach. Define an area where the collection is indicated and scoop/scrape the dust from the surface, using a fiber-free (e.g., a cellophane envelope), contaminant-free scraper (e.g., a stainless-steel spatula). At the laboratory, large samples are homogenized, and a representative sampling of the entire mix will be extracted. For this reason, when deciding how much to collect, the environmental professional may wish to restrict sample sites and limit collection to clearly defined, distinct areas. For instance, an air supply louver in a complaint area and settled dust from a recently cleaned automobile are clearly both defined, well-delineated sample locations.

Then, too, a building material or structural component may appear to be contaminated with an unidentified substance which has become part of its substrate. For instance, a weakened spot on a steel beam may be associated with an unidentified material, or gypsum board may have what appears to be a microbial growth. If possible, collect a piece (e.g., at least a 4 inch square surface) of the substrate. Place it in a plastic baggy and ship with instructions to the lab that you wish to know what the associated material is and describe its physical appearance as best you can. If the instructions are incomplete, the laboratory might just miss the point of the request. Be clear and concise in your instructions!

Outdoor dust sampling is a little more challenging. First, look for smooth or semi-smooth surfaces so one of the methods mentioned previously may be used. Tree leaf surfaces provide an excellent means for collecting ambient dust, and by knowing the time the tree comes into season, leaves provide a time frame for dust deposition. Tops of cars, posts, and window/door frames are other possible sources.

Contaminated snow has been used effectively and provides a generalized time of deposition. Some may even wish to have the particles picked up in raindrops be analyzed. A recent rain measuring collection container may provide a recent example of air particulates picked up by the moisture in the air. Both materials should be handled as one would handle a water sample. Collect a minimum one liter sample.

Textile/Carpet Sampling[20]

Microvacuuming of textiles and carpeting has proven inferior to direct extraction and sonication of textiles and carpeting. See Table 7.4. Microvacuum sampling involves cutting a piece from the textile/carpet a minimum of 100 square centimeters in size. This is placed in a wide-mouth polyethylene jar or zip-lock bag. At the laboratory, the sought after substance is extracted, suspended in water, and filtered through a polycarbonate filter or cellulose ester filter for analysis by transmission electron microscopy. This method is primarily use for asbestos analysis in carpet samples.

Table 7.4 Comparative Sampling Approaches for Asbestos-Contaminated Carpet Analyses

Sample Number	Carpet Piece	Microvac
1	4,800,000	21,000
2	3,300,000	30,000
3	<5,400	<350
4	3,800,000	74,000
5	3,000,000	50,000
6	2,500,000	95,000
7	3,600,000	18,000
8	4,700,000	35,000

< The less than symbol indicates the limit of detection for the method used.

Excerpted from *Methods for the Analysis of Carpet Samples for Asbestos. Environmental Choices.*[21]

Soil Sampling

Soil clean-up is an expensive proposition where contamination has been confirmed. Yet, it is not always clear as to the source of the contamination. Oftentimes, guilt by association is the driving force. Yet, associations may not truly reflect the sequence of events and activities lending to the situation. Confirmation may be as simple as looking to the forensic microscopist for assistance. Some of the more frequently sought soil contaminants which lend themselves more favorably to forensic microscopy include the following:

- Asbestos
- Lead
- Polychlorinated biphenyls

Soil sampling is traditionally performed incrementally to determine in which layer of soil a known or suspect contaminant may be found and quantified. Surface sampling is performed of the uppermost one centimeter of soil with the upper two feet of top soil sampled in six inch increments. The collection is typically performed within a 100 centimeter square area with a minimum yield of 100 grams of soil. Where the top layer is covered with grass or other vegetation and root systems, the collection may require a 5 centimeter depth with a coring device. A stainless steel trowel or similar nonreactive tool should be used to collect the samples.[22]

If there is a possibility that microscopic analysis may be indicated after the initial chemical evaluation, a homogeneous mix from the sample(s) should be saved in a separate 5-milliliter plastic or glass container. Once collected, the samples are generally composited and/or containerized individually and sent to a chemical laboratory for analysis. Initial contaminant confirmation and quantification may be ascertained chemically. Where a contaminant has been confirmed and the material has no obvious or clear source, the homogeneous mix which was saved may be evaluated microscopically for the known contaminant and associated material.

The contaminant may be identified under the microscope through any of a number of techniques (e.g., index of refraction). Its physical appearance and that of associated material will assist the microscopist in determining the probable source of the contaminant.

ANALYTICAL PROCEDURES[23]

Depending upon the microscope of choice, the microscopist has the ability to detect, identify, and measure trace quantities of a substance down to the elemental composition and structural configuration of molecules. Most particles larger than 1 micrometer in size can be identified by visible light microscope analysis. This includes the inspirable mass fraction, thoracic mass fraction, and alveolar respirable mass fraction of the breathing process.

In indoor air quality situations, the environmental professional may consider the particle sizes and the location within the respiratory tract impacted the most where respiratory symptoms are evident. Symptoms impacting only the nose and throat, not the lower portions of the lungs, may be related to particle size (reducing the probability of the smaller diameter particles). This information may be used to determine the type of analytical procedure(s) best suited for a given situation.

Particles that are 1 microns, or less, or never are difficult, if not impossible, by visible light microscope analyses. Under these conditions, special microanalytical procedures may solve the dilemma and/or confirm suspect materials. Each of the various approaches is discussed herein.

Visible Light Microscope Analysis

An experienced microscopist may identify most, not all, sample particulates (greater than 1 microns in size) in a few seconds, if not immediately, without altering the chemical and physical properties of the material. Like differentiating a tree from a light post, when seen and identified on a frequent basis, most microscopic material is easily identified.

Visible Light Microscopy: ***Identification of particles greater than 1 μm in size.***

Parameters which are used in the microscopic analyses include, but are not limited to, the following:

- Size
- Shape
- Color
- Homogeneity
- Transparency
- Magnetic qualities
- Elasticity
- Specific gravity
- Refractive indices
- Birefringence
- Extinction
- Dispersion staining

Sample particles (larger than 1 micron in size) are identified with minimal effort by visible light microscopy. This is done through differences in size, shape, homogeneity, color, transparency, magnetic qualities, and specific gravity. Again, on occasion, additional information is necessary for positive identification of suspect material or for identification of a substance which the microscopist has yet to become familiar with.

The remaining parameters are ascertained through the use of polarized light microscopy which greatly increases particle characterization, and competence. It also ensures positive identification of types of fibers, minerals, and some industrial pollutants.

Each of the above parameters is herein discussed, not to enable the reader to duplicate these methods, but to enable the reader to understand and be conversant in the topic of forensic microscopy.

Size [24]

The first identifying characteristic is particle size. Usually an "estimation" of the average diameter and range is sufficient. However, should greater precision be required, a calibrated micrometer ocular is used.

Sand grains are quite large and variable, whereas sandblasting grains have generally been graded and have a tendency to be more consistent in size. Silica used in cosmetics as an anti-caking agent is around 10 to 20 microns. Used in lacquers, silica is 2 to 3 microns.

Ragweed pollen appears very similar to pigweed pollen with the exception of diameter. Ragweed is approximately 19 microns, and pigweed is 26 microns.

Cocci (round shaped) bacteria tend to be less than 1 micron in diameter, and rounded fungal spores are typically greater than 5 microns. In this case, however, size is not always a definitive factor. A few bacteria/bacterial spores are larger than 2 microns and some fungal spores are as small as 1 microns.

Shape [25]

Shapes vary from highly structured forms (typically biological or man-made) to complex, almost nondescript structures. Descriptive terms and definitions have been developed by the various disciplines (e.g., crystallographers), but they are highly variable and difficult to follow. One source which has attempted to generalize the terms and provides a large encyclopedia of photomicrographs is *The Particle Atlas* (available through McCrone Associates in Chicago, Illinois). Although this atlas is one of the most complete, its coverage is very broad and although extensive, certainly not complete. Experience and an extensive library of knowns are the best sources for easy identification of particles.

Talc is a hydrated magnesium silicate which is typically nonfibrous in nature (fibers undetectable in talc mined from Montana) but has been found to be consist of as much as 50 percent fibers (some New York mines). Used in powders, pharmaceuticals, ceramics, and rubber, talc can not only be identified by its composition, but its association with other materials and its fibrous content. Fibrous content will further dictate its exposure source as well as the mine(s) from which it came.

Silica comes in may diverse forms. Amorphous and crystalline silica can be identified by chemical analysis, but only the microscope can ascertain form. See Table 7.5.

Table 7.5 Forms of Silica

Diatoms	Vitreous silica
Radiolaria	Silica gel
Sponge spicules	Quartz
Silica cells from plants	Sea sand
	Desert sand
Opal	
Flint	Crystobalite
Amethyst	Tridymite

Excerpted from *Microscopical Identification of Atmospheric Particles.*[26]

Tetraethyl lead (from leaded-gasoline exhaust contamination of the ground along roadways) in soils may be differentiated from lead which is associated with paint pigments (where the paint has flaked from a building) or from an industrial smelter (exhaust stack) by its shape, size, and associations with other material. See Figure 7.3.

Many insect parts are readily recognized by microscopy, but species identification requires the whole insect and knowledge in entomology and/or the use of immunoassay analyses of the parts-and-pieces. There are also some chemical tests available to the microscopist whereby chitin may be identified, permitting the microscopist to confirm insect parts-and-pieces only.

Epithelial cells (e.g., human skin cells) generally appear as slightly collapsed cellular material and do not have a nucleus. Skin cells are shed continuously, therefore common in all occupied indoor air environments.

Plant fibers are distinguished by their cell-like structure and absence of scales. Both treated and untreated woods demonstrate pitting along the edges which gives a lacy appearance to the structure. With the exception of feathers and silk, most animal fiber (or hair) is distinguished by the presence of scales. Manmade fibers are perfectly smooth and have continuous filaments.

The form of almost any material may also be instructive as to how it was treated. For instance, wood undergoes predictable, distinct stages when it is incinerated. Wood products (e.g., paper) darken between 150° and 200° C. At about 300° C, the wood material will flash into a flame with the temperature increasing rapidly to 700° to 800° C. At this point, a lacy network of white ash, more or less coherent, develops as the carbon material burns. This charred material is still recognizable as a wood product because it still retains the pitted fibrous structure of the original material. The next change occurs at around 1000° C. The structure collapses (or sinters) and leaves white, glassy spheres of silica which at this stage contain gaseous bubbles. Around 1,100° to 1,200° C, the silica becomes fluid, and the bubbles are released. The final combustion product is a colorless, transparent sphere of silica.[25]

Color[28]

Although best observed in transparent particles, surface color and texture(s) are observable in opaque materials through special manipulation of the light source to obtain top lighting. Thus, all particles can be evaluated for surface appearance, independent of their opacity.

The pigment mercuric arsenate is yellow. Lead chromate is street marker yellow, and lead sulfate is white. Inert phenolic resin dust appears as blue-green granules in cinnamaldehyde-butyrolacetone.

Particles of titanium white, cadmium sulfide, and jeweler's rouge are opaque and appear to be black without top illumination. With the extra light source, however, colors become more apparent. In the preceding three opaque substances, the colors are, respectively, white, orange, and red. Carbon black is, however, black as the name implies.

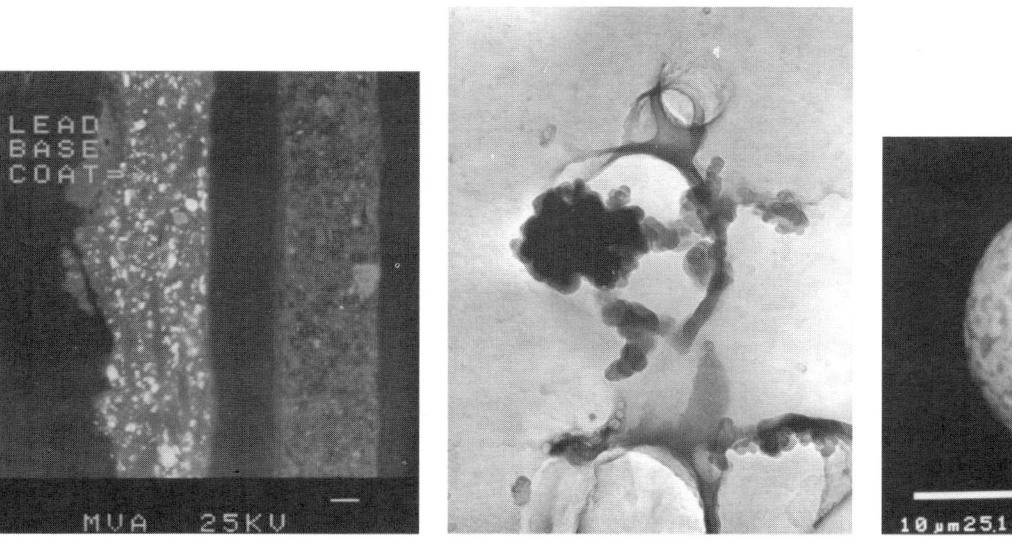

Lead Paint Chip (A)
SEM 450X

Leaded Gasoline Exhaust (B)
TEM, 31,000X

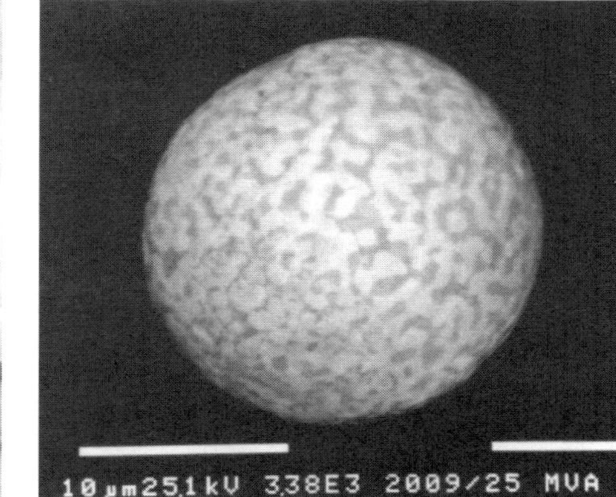

Leaded Solder (C)
SEM, 2,500X

Figure 7.3 Photomicrographs of lead-containing material. The first photo (left) shows layers of paint, each less than 0.01 inch thick, magnified 450x by scanning electron microscope in backscatter mode; the bright spots depict particles which contain lead. The second photo (middle) shows particles collected from within the Prague Tunnel (Czech Republic) where leaded-gasoline still is used, magnified 31,000x by transmission electron microscope. The third photo (right) shows a 20 micron sphere of lead-tin solder, magnified 2,500x by scanning electron microscope in backscatter mode. (Courtesy of MVA, Inc., Norcross, GA)

Homogeneity [29]

Homogeneity is the particle association or lack of associations with other materials and/or particles. In some instances, the only way to differentiate synthetics from naturally-produced products is through observing their relationship with other substances. Synthetics will tend to be pure (or homogeneous), whereas natural products are generally associated with other materials. Some natural products even provide hints as to the area soils from which them came.

Magnetite spheres are often found in atmospheric dust and have occasionally been identified as "micrometeorites" unless associated with transparent spheres of pulverized coal boiler flyash. In the latter case, the magnetite is likely to be from a power plant stack, whereas the combustion products are iron minerals associated with pulverized coal.

A thin brown coating on the surface of rounded quartz or other sand grains is an indication of spent molding sand which was used in a foundry. The coating is the bonding material typically used to hold the sand grains for metal-casting molds.

Transparency [30]

A transparent substance is one which transmits all or nearly all of the light entering the material. The more transparent a substance (e.g., glass and asbestos), the more difficult it may be to see under the microscope without special lighting, particle discontinuities, natural pigmentation by impurities, or special staining procedures.

Special lighting may involve phase contract microscopy. Particle discontinuities include, but are not limited to, air bubbles and variability of the internal structure (e.g., paraffin crystals which have random orientations and allow light to pass at varying angles). Natural pigmentation by impurities, depending upon the amount of unrelated material and its opacity, may render color to a transparent substance in the same fashion that impurities in the air lend color to the sky. Staining procedures are discussed later.

An asbestos fiber is transparent versus an opaque wood fiber. Gypsum becomes transparent after being heated to temperatures in excess of 100° C and upon dehydration, at which time, the gypsum becomes plaster of paris. The basic chemistry remains the same with the exclusion of water, yet there is a distinct difference in material. Whereas pure glass is transparent, iron impurities impart a greenish color to the material. Impurities also give color to and devalue gem-quality diamonds. Oil refinery catalysts, although transparent, have a random matrix which causes them appear optically discontinuous (like paraffin wax).

Magnetic Qualities [31]

Suspended particles in a liquid drop on a microscope slide may be observed for responsiveness to a micro magnet. Isolation of magnetic metals and exotic magnetized ceramic materials may be accomplished easily, and these materials may be withdrawn from the sample to be analyzed separately. See Table 7.6.

Table 7.6 Magnetic Materials

Metals and Their Compounds	Ceramics
Iron	Ferromagnetic ceramics
Iron oxides	Ferrites
Iron-silicon alloys	Barium ferrite
Nickel irons	Strontium ferrite
Chromium steels	
Tungsten steels	
Cobalt steels	

Excerpted from *Materials Handbook.* [32]

The suspect material is suspended in a viscous liquid (e.g., melted Aroclor) and subjected to a miniaturized magnet. Those particles which are magnetic will respond to the movement of the magnet which is manipulated by the microscopist.

Elasticity and Hardness [33]

Elasticity and hardness are determined by the "squoosh" test. A particle is manipulated within a viscous liquid and subjected to pressure. Elastomers will not only flatten as pressure is applied, but they will recover to their original shape when the pressure is released, not unlike an automobile tire. Soft materials that have no elasticity will flatten and retain their flatness when the pressure is released. Hard structures will crack, and many will cleave or split along a defined, natural line.

An elastomers, according to the American Society for Testing and Materials, is "a polymeric material which at room temperature can be stretched to at least twice its original length and upon immediate release of the stress will return quickly to approximately its original length." This includes natural rubber which is extracted from the sap of rubber trees, synthetic rubber, and synthetic elastomers (e.g., neoprene, nitrile elastomers, silicone elastomers, etc.). Most other soft materials can be excluded from the list.

Specific Gravity[34]

Generally, specific gravity is determined along with the refractive index determination which requires polarized light microscopy. If ascertained singularly, however, only a biological microscope (commonly referred to as a "light microscope") is required.

An estimate of the density is performed based upon experience and typical ranges for a given type of material. Hairs and fibers range from 1 to 2. Soil has a broader range of gradients. See Table 7.7 for a general overview of specific gravity for a variety of unknowns.

Table 7.7 Specific Gravity of Materials

Substance	Specific Gravity
Wood	
Cedar, dry	0.36
White pine, dry	0.41
Redwood, dry	0.42
Fir, dry	0.56
Ash, dry	0.63
Maple, dry	0.65
Tar	1.00
PCBs	1.182-1.-812
Magnesium	1.74
Beryllium	1.848
Concrete	2.3
Glass	2.5
Limestone	2.5
Asbestos	2.3-2.6
Graphite	2-3
Aluminum	2.7
Portland cement	3.0
Titanium	4.8
Chromium	7.14
Cast iron	7.2
Steel	7.8
Copper	8.94
Lead	11.38
Platinum	>20.0

Excerpted from *Materials Handbook.*[35]

Upon estimating the approximate density of an unknown substance, the microscopist chooses a chemical from a broad spectrum of liquids with known densities, places it on a microscope slide, and suspends the unknown substance or group of substances in the drop for observation. If the material being tested goes to the top of the drop, it is lighter than the liquid. If it goes to the bottom, it is heavier than the liquid. When it remains in suspension, floating freely within the chemical, the specific gravity of the unknown approximates that of the liquid in which it is suspended.

Table 7.8 Properties of Some Organic Liquids

Substance	Density	Refractive Index
Kerosene	0.8200	—
o-Xylene	0.8745	1.508
Chlorobenzene	1.1066	1.525
Nitrobenzene	1.1987	1.553
1,4-Dichlorobenzene	1.2880	1.546
Bromobenzene	1.4991	1.560
1-Iodonaphthalene	1.7344	1.705
Iodobenzene	1.8320	1.621
1,4-Dibromobenzene	2.2610	1.574
Bromoform	2.8900	1.598
Tetrabromoethane	2.9640	1.638
Methylene Iodide	3.3250	1.730

Excerpted from Density and Refractive Index—Their Application in Chemical Identification. [36]

The choice of test liquid is also important. The test liquid should have little or no effect on the specimen being examined, and it should evaporate off to allow for other test liquids to be tried. The chemical should not cause the unknown substance to dissolve, swell, or disintegrate. Several attempts may be required to obtain the needed information. Material density data contributes significantly toward the identification of unknowns. Some of the more commonly use organic liquids with known densities are listed in Table 7.8.

Refractive Indices [37]

Refractive index is the ratio of the velocity of light in a vacuum to the velocity of light through a substance. It always exceeds 1.00, and the higher the refractive index—the slower light passes through the material. Thus, as the density of a substance increases, so does the refractive index. The refractive index for water is 1.33.

The refractive indices are measured with the aid of a single-direction (i.e., east-west) polarized light microscope and mounting liquids of known refractive indices. The polarizer vibration is arranged east-and-west with respect to the field of view. Then, by rotating the microscope stage, any particle in the field of view may be manipulated, positioned to its vibration direction(s).

Isotropic particles have the same optical properties no matter what orientation they are placed in. This includes cube-shaped crystals and all noncrystalline substances which are not internally stained (e.g., glass wool fibers). An example of an internally stained noncrystalline substance is organic, polymeric fibers (e.g., Nylon), and they are not isotropic. In brief, isotropic particles have a single refractive index, "regardless of their orientation," for a specified temperature and a given wavelength.

Anisotropic particles poses different optical properties, refractive indices as their orientation changes. This includes noncube-shaped crystals (e.g., triangles, rectangles, and tetragonals) and stainable noncrystalline substances (e.g., polymeric films). There are at least two different refractive indexes for anisotropic particle, depending upon the particle orientation. As observed under the microscope, the refractive index of the particle in view is always oriented on the east-west plane. So, anisotropic particles must also be oriented in the other direction, or directions, and refractive index determined for each orientation.

Once properly oriented, the difference in light absorption between the unknown and the mounting solution is roughly related to the difference in refractive indices for different directions in the particle. Once again, as with specific gravity determinations, the microscopist guesses at the refractive index, then tests the theory. The closer the refractive index of the unknown substance is to the known index of the mounting solution, the more diminished is the boundary contrast.

The microscopist can determine if the refractive index of the unknown is higher or lower than that of the liquid by observing that which is referred to as the Becke line. As the microscopist raises the focal plane of the microscope, the Becke line (which appears as a halo) moves toward the medium of higher refractive index. If the halo moves closer to the particle, the particle has the higher refractive index. If it moves away from the particle, the particle has a lower refractive index. The converse is true whenever the focal plane is diminished.

As the refractive index of the unknown particle approaches that of the mounting solution, it becomes more invisible. The greater the difference, the greater the contrast, and the unknown becomes easier to see. In this fashion, clarifying the refractive index of an unknown helps narrow the gap between unknown and known.

An excellent example of materials identification and differentiation through refractive index determination is that of synthetic fibers. The refractive index of undyed polyester Dacron fibers is 1.711 ± 0.011 when observed in parallel polar alignment and 1.541 ± 0.012 when observed in perpendicular polar orientation, whereas the refractive index dyed Dacron polyester fibers is

1.714 ± 0.014 and 1.748 ± 0.013, respectively. The refractive indices for Kevlar fibers exceed 2.0 on both alignments, and asbestos is between that of polyester and Kevlar fibers.

The refractive index for isotropic crystals of sodium fluoride is 1.326, and that of diamonds is 2.42. The refractive indices for more complex structures may involve as many as three orientations. For instance, borax has refractive indices of 1.447 (alpha orientation), 1.469 (beta orientation), and 1.472 (gamma orientation). Sulfur is at the high end of the scale with 1.950 (alpha orientation), 2.038 (beta orientation), and 2.241 (gamma orientation). The high end of an intermediate structure is hematite with refractive indices of 3.22 (omega orientation) and 2.94 (epsilon orientation).

Birefringence [38]

All anisotropic substances are said to show birefringence. The numeric difference between the highest and lowest refractive indices of a substance is referred to as birefringence. The birefringence of hematite, whose refractive indices are mentioned above, is calculated to be 0.28. Yet, determination of refractive indices, particularly for the anisotropic substances can be tedious and time-consuming. So, there is a short-cut!

Birefringent particles have distinct, consistent colors when observed through a polarizing microscope. The resultant colors are based upon particle thickness and light retardation. Thickness can be physically measured, and light retardation can be observed in the form of color and brightness.

One of the principal difficulties this method, however, lies in measuring particle thickness as the particle must at times be reoriented (bounced, giggled, or micromanipulated) in order to measure the vertical distance parallel to the light path. Fibers are, by far, the easiest the measure. Crystals which have as many as three refractive indices (or three orientations) and are not structurally distinct are a little more elusive to orient. Re-orientation for thickness measurements of complicated crystalline structures is truly an art-unto-itself!

To determine color and brightness to within the proper order (or brightness), a compensator is used (the explanation not to be disclosed within this book). With the known particle thickness and determined retardation, the microscopist then refers to the Michel-Levy Birefringence Chart (which was published in 1888 and is still in use). From the chart, birefringence is easily determined.

The birefringence for any two-dimensional fiber or three-dimensional crystal is then matched with that of known substances. Some typical substances are also included on the chart—for quick reference. This approach confirms, greatly facilitates, and simplifies the identification of minerals and fibrous structures.

Extinction [39]

Crystals and fibers may be oriented such that their polarization colors, when the microscope stage is rotated, are observed as black, disappearing into the dark background. These four positions of darkness for anisotropic particles are referred to as "extinction positions." Most crystalline substances and fibers show uniform extinction. This is where all portions of the particle disappear at the same time (e.g., fibers that are lying perfectly straight, parallel to the surface of the microscope slide).

There at three types of extinction. They are parallel, symmetrical, and oblique. Parallel is where the vibration directions are parallel to the edges. Symmetrical is where opposing sides are the same. Oblique is where the edges vary, deviate from one another.

The particle is aligned in the center of a rotating stage with measurable angles. With a prominent edge centered and parallel to a cross line, the stage is rotated until the edge shows uniform extinction and disappears, taking on a ghost image. The difference (which never exceeds 45 degrees) in the start angle and the angle at which the edge disappears is the "extinction angle."

This approach is used primarily to identify and confirm soil minerals and crystalline structures. Extinction positions are either published and/or recorded by the individual microscopist in a lab journal.

Dispersion Staining [40]

Dispersion staining is a means of particle identification whereby particle identification is based on the difference between refractive index dispersion of a particle and its liquid medium. Two different procedures are based on the use of stops in the objective back focal plane, and they are observed as either a specific color "boundary" or its opposing color. Information regarding the anticipated colors can be obtained from charts of known materials as observed in various liquid immersions, at different wavelengths. These are in the form of a curve for each particle, plotted with liquid refractive indices to the wavelength anticipated..

This process makes it possible to systematically identify particles by the colors observed when immersed in a liquid of known refractive index. Its most common use in the environmental profession has been for identification of asbestos types. Other uses have been as follows:

- Determination of toxic dusts
- Identification of settled dust particles
- Synthetic fiber identification
- Mineral characterization

Some of the more obscure uses included:

- Penicillin G in an air sample from a sulfa drug manufacturing area
- Quartz in mine atmospheres
- Heroine dust vacuumed from an automobile

In the hands of an experienced microscopist, this method is fast and effective. It detects traces of materials down to parts per million in any given mixture. There is, however, one important limitation!

Dispersion staining is reliable for confirmation of suspect material only. Where an unknown is involved, this method cannot be used singularly without a reasonable suspicion. The other methods mentioned above and in the succeeding sections can be used in combination with dispersion staining to effect a reasonable conclusion.

Specialized Microscopic Techniques

Specialized techniques are available for use when the particle size is less than one micron, when the price is not a consideration, or for confirmation in litigation or high visibility cases. Occasionally, other techniques become necessary when dealing with extremely small particles or exotic mixtures.

X-ray Diffraction[41]

Other than visible light microscopy, X-ray diffraction is the only other technique available which permits identification and differentiation of crystals. Where there are three different forms of silica (i.e., quartz, tridymite, and cristobalite), chemical analysis may confirm the presence of silica but not be able to differentiate the type.

The sensitivity is down to approximately 10^{-2} nanograms, and the procedure is nondestructive of the sample. Although it may be as small as one particle, measuring 5 microns in diameter, the ideal sample size is 40 to 50 microns. The smaller particles require more extensive manipulation (e.g., removal of air from within the camera), so the cost for analysis may increase with smaller particle sizes.

X-ray diffraction measures the interplanar spacing of atoms in a crystal. These spacings are unique for every compound and are identified by comparison with prior tabulated known compounds. This comparison is performed with the assistance of a computer file which has well over 20,000 substances in its data bank. The data file is constantly being expanded. If a sample is suspect of containing a specific substance which is not on file, the known substance (in its pure form) may be scanned and entered into the data banks to be used as a reference for the unknown.

Scanning Electron Microscope [42]

Scanning electron microscopy (SEM) comes into play where a sample size is too small to be observed by visible light microscopy (equal to or less than 1 micron in diameter) or where greater resolution and depth of field of larger particles (from 1 to 100 microns in diameter) is required. It operates in a similar fashion to that of the stereo binocular microscope, refracting electron beams (instead of visible light) off the surface of a sample. These refracted electrons are projected onto a viewing camera or film to permit the analyst to observe the structure(s).

Scanning Electron Microscopy:
Morphology, spacial, and inorganic elemental analysis of particles down 0.2 μm in size.

The SEM is capable of magnification of particles typically around 0.2 micron in diameter. Where the depth of field for visible light microscopy is around 1 micron, it is 300 micron for the SEM. This allows for greater contrast, ease of viewing the unknown sample.

The resolution is about 300,000 times the actual particle size, or 200 times greater than that of the most powerful light microscope (which has a magnification capability of 1,500 times). Smaller particles are more readily observed due to the increased magnification. This is generally the case with metal fumes, clays, some pigments, bacteria, and viruses.

Another added feature to the SEM is the ability to add an energy dispersive X-ray analyzer (EDXRA) to the unit. This X-ray analyzer is capable of greater detection than that of X-ray diffraction. Where the X-ray diffraction provides a means for identifying compounds, the EDXRA can detect elements (above nitrogen on the periodic table). Most analysts agree that without this added ability for detecting elements, SEM would be inferior to visible light microscopy in its detection ability.

Still spores, bacteria, and viruses can be identified as spores, bacteria, and viruses only. Viruses and bacteria are generally smaller than 1 micron in diameter and can be identified as such only through the use of SEM due to the higher resolution. To type these biologicals by genus and species, the sample must still be cultured or manipulated in some other fashion besides microscopy.

Table 7.9 Most Common Elements of the Earth in Order of Decreasing Abundance

Oxygen
Silicon
Aluminum
Iron
Calcium
Magnesium
Sodium
Potassium
Titanium
Hydrogen
Carbon
Phosphorous
Manganese
Sulfur

Particles greater than 1 micron in diameter may still require the EDXRA for identification and are frequently more easily identified by visible light microscopy. Thus, SEM with EDXRA may be used as a secondary means of identification for the larger particles, and as the primary means of analysis for particles at or less than 1 micron in diameter.

Transmission Electron Microscope[43]

Transmission electron microscopy (TEM) analysis works in a similar fashion to that of the biological microscope by penetrating a sample with focused electron beams instead of visible light. These electron beams are observed in a similar fashion to that of SEM where the beams are projected onto a viewing screen or film.

> **Transmission Electron Microscopy:**
> ***Identification and product analysis of particle/components down 0.5 x 10^{-3} μm in size.***

The depth of focus is 1 micron and its resolution is 0.5 x 10^{-3} micron. The maximum prepared particle thickness is 0.05 micron, and the maximum sample diameter is 3 millimeters. The TEM can be fitted for selected area electron diffraction (SAED) and EDXRA. The SAED functions in a similar fashion to that of X-ray diffraction limiting the coverage area, and the electron beam is used to measure the interplanar spacing of atoms in a given area. The SAED information is compared with a data bank for compound identification, and the EDXRA provides the elemental fingerprint.

Particles are scanned for structural appearance, compound identification, and elemental fingerprinting. If a search is performed for a specific substance, the microscopist reports results in percent by weight. It is important to note that where asbestos content is performed by polarized light microscopy, the results are provided as percent by volume where the definition of asbestos is a "mixture containing greater than 1 percent by weight " of certain types. The only means available to provide true "percent by weight" is through TEM.

Electron Microprobe Analyzer [44]

The electron microprobe analyzer (EMA) is an ultramicroanalytical tool which can be used to enhance a light microscope, SEM, X-ray fluorescence, and cathodoluminescence. It is, also, referred to as "mass scanning."

A sample containing a large number of small particles may be rapidly characterized by chemical composition. This is generally performed by automation of the specimen stage, scanning beam, and spectrometer. In this case, a few thousand particles can be characterized from any given sample.

The electron microprobe analyzer may be used to locate a "needle-in-a-haystack." If searching for a known substance which may be present in a sample in only parts per million, or trace levels (e.g., asbestos fibers in urban air), the analyzer is ideal. It is set up to identify an element, or combination of elements, which are present in the substance of concern. Each time the substance is located, the stage stops, and that particle is quantitatively analyzed. Then, the stage continues its search. The ideal lower limit for adequate identification is 0.1 percent, but the method is capable of locating down to 10^{-4} percent. The latter involves a considerable amount of time consumption, therefore, cost, but it is possible.

Samples as small as 1 micron in diameter can be analyzed for most elements present in the sample to 1 percent or greater. This constitutes a detection limit as low as 10^{-4} nanograms. An electron beam is focused to a spot smaller than 1 microns square in area. The characteristic X-rays emitted from the spot are analyzed for wavelength, or energy, dispersing systems both quantitatively and qualitatively. A limitation is its inability to detect lithium (sometimes beryllium), and one hundred parts per million is the lower limit of detection for most elements.

The analyzer is capable of mass scanning of between 4 and 50 particles per hour. The speed and ease of analysis allows for any given sample, containing "up to 1,000 unknowns as little as 20 hours to analyze"—a 24-hour turnaround, in a pinch!

Ion Microprobe Analyzer[45]

The ion microprobe analyzer (IMA) provides a means for mass spectrometry on small particles or small areas of "bulk" samples. This method is one of the most sensitive tools available for small particle analysis. It is sensitive to every element in the periodic table and can, under ideal conditions, detect as little as 10^{-20} grams of some elements, 10^{-19} grams of most elements. It is fully capable of analysis of trace amounts of material from samples as small as 1 micron and, in some instances, can obtain parts per billion of some elements. The time required for semiquantitative analysis is typically 40 seconds, as short as 4 seconds.

The instrumentation consists of a light microscope (which is used to locate the sample), an ion source, a column of two electrostatic lenses, and a mass spectrometer.

The versatility of the IMA is similar to that of the electron microprobe analyzer, yet it is much faster and can assess particles which are much smaller (e.g., 1 part per billion instead of 1 part per million). Any airborne, waterborne, or contaminant particles can be analyzed with this tool. A few examples include the analysis of micrometeorites, lead particles from auto exhaust, and contaminants on integrated circuits.

COMMERCIAL LABORATORIES

Due to the extensive training required to be proficient in all aspects of this field, there are a limited number of laboratories capable of responding to all the nuances which may arise in an environmental evaluation. An experienced microscopist may cost a little more per hour yet be able to provide results with less time expenditure than one with less experience and lower rates.

On the other hand, the desired information may be obvious (e.g., heavy concentrations of ragweed pollen) and readily apparent to even the inexperienced microscopist who may serve as an initial previewer. Many of the commercial laboratories are accustomed to analyzing primarily for asbestos only. Quarry the commercial laboratory as to its capabilities and limitations. Those experienced in performing forensic analyses can easily apply the forensic knowledge to environmental issues.

REFERENCES

1 McCrone, Walter C. The Shroud of Turin: Blood or Artist's Pigment? Acc. Cehm. Res. 23:77-83 (1990).

2 McCrone, Walter C. Experiences with the Shroud of Turin. [Oral communication] Forensic Microscopy Course, McCrone Research Institute, Chicago, Illinois, November 1993.

3 Bisbing, Richard. Clues in the Dust. *American Laboratory*. [Reprint] November 1989.

4 Ibid.

5 McCrone, Walter C. and Skip Palenik. The Solids We Breath. *Industrial Research*. April 1977. p. 66.

6 McCrone, Walter C. Environmental Pollution Analysis. *American Laboratory*. (December 1971)

7 Ibid.

8 McCrone, Walter C. and Skip Palenik. The Solids We Breath. *Industrial Research*. April 1977. p. 66.

9 McCrone, Walter C. Microscopy and Pollution Analysis. *American Laboratory*. (July 1970)

10 McCrone, Walter C. Forensic Microscopy—Appendices: [Training Manual for Forensic Microscopy] McCrone Research Institute, Chicago, Illinois, October 1989.

11 McCrone, Walter C. Detection and Measurement with the Microscope. *American Laboratory*. [Reprint] December 1972.

12 Hedge, A., W.A. Erickson and G. Rubin "Effects of Man-Made Mineral Fibers in Settled Dust on Sick Building Syndrome in Air-Conditioned Offices." Proceedings from a Conference on Indoor Air, 1993.

13 McCrone, Walter C. The Solids We Breathe. *Industrial Research*. April 1977.

14 McCrone, Walter C. Microscopy and Pollution Analysis. Reprint from "Measuring, Monitoring, and Surveillance of Air Pollution," *Air Pollution*. Volume III (1976).

15 Bisbing, Richard E. Microscope and Pollution Analysis. [Oral communication] McCrone Associates, Inc., Chicago, IL. (June 1995)

16 McCrone, Walter C. *Air Pollution*. Academic Press, Inc., New York, New York, 3rd Edition, Volume III, 1976. pp. 101-2.

17 Millette, J.R., et. al. Scanning Electron Microscopy of Post-it Notes Used for Environmental Sampling. *NAC Journal*. Spring:31-35 (1991).

18 McCrone, Walter C. *Air Pollution*. Academic Press, Inc., New York, New York, 3rd Edition, Volume III, 1976. pp. 102.

19 Millette, J.R., T. Kremer, and R.K. Wheeles. Settled Dust Analysis Used in Assessment of Buildings Containing Asbestos. [Bulletin] McCrone Environmental Services, Inc. Norcross, Georgia, 1990. pp. 216-219.
20 Millipore Product Literature. Millipore Corporation, Bedford, Massachusetts, 1996.
21 Millette, James R., et. al. Methods for the Analysis of Carpet Samples for Asbestos. Environmental Choices Technical Supplement, March/April 1993.
22 Ness, Shirley. *Air Monitoring for Toxic Exposures*. Van Nostrand Reinhold, New York, New York, 1991. p. 414.
23 McCrone, Walter C. Forensic Microscopy—Appendices: [Training Manual for Forensic Microscopy] McCrone Research Institute, Chicago, Illinois, October 1989.
24 Ibid. pp. 6.6-6.7
25 Ibid. pp. 6.8-6.9
26 McCrone, Walter C. Microscopical Identification of Atmospheric Particles. [Bulletin] McCrone Institute, Chicago, Illinois.
27 McCrone, Walter C. and John Gustav Delly: *The Particle Atlas*. Ann Arbor Science Publishers, Inc., Ann Arbor, Michigan, 1973. p. 323.
28 McCrone, Walter C. Forensic Microscopy—Appendices: [Training Manual for Forensic Microscopy] McCrone Research Institute, Chicago, Illinois, October 1989. p. 6.11.
29 Ibid. pp. 6.11-6.12
30 Ibid. pp. 8.5-7
31 Ibid. p. 10.18
32 Brady, George S. and Henry R. Clauser. *Materials Handbook.* McGraw-Hill, Inc., New York, New York. pp. 505-9
33 McCrone, Walter C. Forensic Microscopy—Appendices. [Training Manual for Forensic Microscopy] McCrone Research Institute, Chicago, Illinois, October 1989. p. 10.17.
34 Ibid. p.
35 Brady, George S. and Henry R. Clauser. *Materials Handbook.* McGraw-Hill, Inc., New York, New York. p. 973.
36 Kirk, Paul L. Density and Refractive Index—Their Application in Chemical Identification. Charles C. Thomas Publisher, Springfield, Illinois, 1951.
37 McCrone, Walter C. Forensic Microscopy—Appendices: [Training Manual for Forensic Microscopy] McCrone Research Institute, Chicago, Illinois, October 1989. pp. 8.7-8.8.

38 Ibid. pp. 10.8-10.14.
39 Ibid. pp. 10.5-10.7.
40 Ibid. pp. 8.9-8.17.
41 McCrone, Walter C. *Air Pollution.* Academic Press, Inc., New York, New York, 3rd Edition, Volume III, 1976. pp. 114-115.
42 Ibid. pp. 118-121.
43 McCrone, Walter C. *Air Pollution.* Academic Press, Inc., New York, New York, 3rd Edition, Volume III, 1976. pp. 121-132.
44 Ibid. pp.132-138.
45 Ibid. pp. 138-143.

Chapter 8

Volatile Organic Compounds

When sampling for unknowns, organic compounds are part-and-parcel of the overall picture. Identification and quantitation of organics have become a well-spun web of concern in numerous situations. Some quarries invoke a duty to identify toxic organic components in the air. Others summon a need to identify respiratory and/or skin irritants. Carcinogenic organics are another issue.

In 1995, the National Institute for Occupational Safety and Health (NIOSH) reported 17 percent of all their indoor air quality surveys have identified volatile organics as either the cause or contributor to the indoor air quality complaints. Many offices spaces have residual organic components in the air from construction, renovation, maintenance, janitorial, chemical usage/processing (e.g., spray painting associated with marketing projects) and pest control activities. There is also off-gassing from new furnishings, building materials, and office supplies/equipment. Some of the organics may originate from the growth of microbes. Tobacco smoke, deodorants, and perfumes contribute to the total organic loading. Some chemicals can diffuse inside from the outdoor environment.

The outdoor environment has different sources of organic compounds. Automobile exhausts and industrial pollutants prevail predominately around large cities or industrial operations. Even food manufacturing operations have been known to generate organic chemicals. Environmental organic compounds also evolve from nature's store of plant life.

Industrial activities generate organic air pollutants both inside and outside. Although most of these chemicals are known, some are by-products of multiple-chemical processing and chemical treatments. Stack exhausts may service several areas with different chemical contributions, and complex chemical reactions in the stacks result in complex chemical mixtures in ambient air.

Incidents of chemical storage fires or petrochemical explosions result in the release of unknown organic by-products. Fires in transport systems and buildings result in the release of unknowns. The possibilities are infinite!

Yet, with all things considered, the sampling and analytical approaches are highly variable as presented by researchers, laboratories, and environmental professionals. Sampling may be performed for total volatile organics according to OSHA and NIOSH procedures, and component identification is accomplished by one of several EPA methods, NIOSH protocols, or special laboratory procedures. There is no clear consensus as to a single best approach.

In defense of the multiple analytical approaches, one protocol cannot possibly apply to all situations. Space age technology has yet to conquer this dilemma. The direction in which most environmental professionals embark, at this time, is to identify as many volatile organics as might be expedient, extend the limits of present technology, and then to target other chemicals anticipated in a given environment.

The intent herein is to present published approaches as well as some of the more common adaptations of published methodologies. These are discussed along with the various nuances to sampling strategies, screening procedures, sampling methodologies, analytical procedures, and interpretation of results.

PUBLISHED SAMPLING AND ANALYTICAL METHODS

There are numerous approaches to sampling for the identification of volatile organics. Some are published. Others are either situation dependent or being developed by researchers. Some of the more commonly used published methodologies are those which have been created by the Environmental Protection Agency. To their credit are the "Toxic Organic Series," referred to as the EPA "TO" Methods. A summary breakdown of the various methods, sampling methodologies, and detection limits is presented in Table 8.1.

TO Methods 1 and 2 are associated with specific compounds, and the search is limited to specific compounds which are summarized in Table 8.2. While the search is limited, results are easier to quantitate through comparison with a standard. Then, too, there is an alternative that covers all bases. This alternative is Method 3.

TO Method 3 covers a search for known as well as a search for other significant peaks. Results typically include the listing of chemicals sought under the method, and the level of detection or actual amount of material detected.

In addition to the Toxic Organic Series, the Environmental Protection Agency has also published some methods for indoor pollutants, referred to as the EPA "IP" Series. These are summarized in Table 8.1. The IP-1B Method for Volatile Organic Compounds neither restricts nor mandates the search for any specific compounds. The method is, however, limited to the capture and analysis of only nonpolar volatile organics.

There are several variations of these EPA methodologies generated by laboratories and individual environmental professionals. Some laboratories restrict their efforts only to established, published methodologies. Others go the extra mile to accommodate special needs.

Table 8.1 Environmental Protection Agency Air Monitoring Methodologies

Methods	Target Analytes	Collection Material	Flow Rate(s)	Maximum Air Sample Volume	Detection Limit	Analytical Method(s)
TOXIC ORGANICS						
TO-1	40 Nonpolar volatile organics	Tenax thermo-packed	6-500 ml/min.	10-250 liters	0.002 μg report: ppb	TD/GC/MS
TO-2	TO-1 Organics + 25 nonpolar "highly" volatile organics	Carbon molecular sieve thermo-packed	15-400 ml/min.	20-100 liters	0.002 μg report: ppb	TD/GC/MS
TO-3	TO-2 + chemical library search	Carbon molecular sieve thermo-packed	100-200 ml/min.	100 liters	0.002 μg report: ppb	TD/GC/MS
TO-4	Organochlorine pesticides + PCBs	Glass fiber filter + PUF*	200-280 l/min.	40 x 10^4 liters	ng/m^3	GC/ECD
TO-5	Aldehydes & ketones	Dinitrophenylhydrazine (DNPH) solution**	100-1,000 ml/min.	80 liters	ppb	HPLC
TO-6	Phosgene	2% Aniline in toluene***	100-1,000 ml/min.	50 liters	< ppb	HPLC
TO-7	Amines	Thermosorb/N tube	100-2,000 ml/min.	300 liters	1μg/m3	GC/MS
TO-8	Phenol & cresols	Sodium hydroxide solution	100-1,000 ml/min.	80 liters	1 ppb	HPLC
TO-10	Organochlorine pesticides	PUF	1-5 l/min.	5,000 liters	0.01-0.1 μg/m^3	GC/ECD (other detectors)
TO-11	Aldehydes & ketones	DNPH Treated-silica gel sorbent	200 ml/min.	300 liters	1-2 ppb	HPLC

Table 8.1 (continued)

TO-12	Nonmethane organics	Ambient air canister	NA	NA	1 ppb	GC/FID (cryogenic treatment)
TO-13	Polynuclear aromatic hydrocarbons	Quartz fiber filter + PUF (ot XAD-2)	200-280 l/min	325 x 10^3 liters	0.1-1.0 µg/m3	GC/FID, GC/MS, or HPLC
TO-14	41 Volatile organics	Ambient air canister	NA	NA	0.1 ppb	GC/MS (cryogenic treatment)
TO-15	Polar & water soluble volatile organics					
TO-16	Broad spectrum of organics and inorganics	Direct reading field/lab instrument (open pass through 1/4 mile or high volume chamber sampling)	NA	NA	Variable	FTIR
TO-17	Volatile organic compounds	Multibed sorbent	—	—	—	GC/MS
INDOOR POLLUTANTS						
IP-1B	Volatile, non-polar organics	Tenax thermo-packed sorbent	6-500 ml/min	20-200 liters	0.002 µg reported in ppb	TD/GC/MS
IP-2A	Nicotine	XAD-4 sorbent	1 l/min	480 liters	0.02 µg/m^3	GC/NSD
IP-6A	Formaldehyde & other aldehydes ****	DNPH-treated silica gel sorbent	200 ml/min	300 liters	1-2 ppb	HPLC

Table 8.1 (continued)

Methods	Target Analytes	Collection Material	Flow Rate(s)	Maximum Air Sample Volume	Detection Limit	Analytical Method(s)
IP-6C	Formaldehyde & other aldehydes	DNPH treated passive monitor	NA	NA	Variable	GC/NSD
IP-7	Polynuclear aromatic hydrocarbons	Quartz fiber filter + PUF (or XAD-2 adsorbent)	20 l/min	30 x 103	< 1 μg/m^3	GC/FID, GC/MS, or HPLC
IP-8	Organochlorine & other pesticides	PUF	1-5 l/min	5,000 liters	< 1 μg/m^3	GC/ECD (other detectors)

* A PUF is a 3 inch by 6 cm diameter polyurethane foam plug (PUF).
** Iced impingers which contain a two-phase mixture of aqueous DNPH and iso-octane.
*** Impinger with solution of 2% aniline in toluene.
**** May also use Waters Sep-Pak silica gel cartridge at a flow rate of 2 l/min, air volume of 1,000 liters, detection limit of < 1ppb.
TO Method 17 will eventually replace TO Methods 1 and 2.
NA Not Applicable
— Information Unknown/Unpublished

Analytical Procedures
GC Gas chromatography
TD/GC/MS Thermo Desorption/Gas Chromatography/Mass Spectrometry
GC/MSGas Chromatography/Mass Spectrometry
GC/FID Gas Chromatorgraphy/Flame ionization Detector
GC/ECD Gas Chromatography/Electron Capture Detector
GC/NSD Gas Chromatography/Nitrogen Selective Detector
HPLC High Pressure Liquid Chromatography
FTIR Fourier Transform Infrared Spectrometry

restrict their efforts only to established, published methodologies. Others go the extra mile to accommodate special needs.

The National Institute for Occupational Health and Safety is in the process of developing an approach as well. It is referred to as the Methodology for the Determination of Selected Volatile Organic Compounds in Indoor Air, and the approach is summarized as follows:[1]

- Target analytes—Volatile organics typically found in indoor air quality studies
- Collection material—Multiple sorbent thermal desorption tube
- Flow rate—less than 200 milliliters per minute (ml/min.) [suggested by prior sampling techniques]
- Sample volumes—1 to 6 liters
- Detection limit—parts per billion (ppb)
- Analytical method—Thermal desorption with GC/MS

Their approach has led to a listing of twenty compounds commonly identified in indoor air quality surveys. See Table 8.3 for a listing of substances implicated, ACGIH limits, and exposure sources.

NIOSH has also performed sorbent and sample placement studies. These are discussed in greater depth within the subsection on Solid Sorbent Sampling.

SAMPLING STRATEGY

When sampling for unknown chemicals, there may or may not be a known source. In a case involving industrial operations, the source is typically known (e.g., chemicals evolved during heat generating machine activities while using a coolant). Where indoor air quality is a concern, the causative source is generally unknown. In the latter case, the focal point may be the general area, central to where complaints have occurred. After determining the source or general area to be sampled, another location may be chosen for background comparison sampling.

In an industrial operation where the source is known, at least one sample area should be identified to serve as a control. This may be a nonwork or nonprocess associated location (e.g., a product packaging area or office space). It may involve before and after sampling where ventilation controls are being tested or where process start-up has yet to begin. The control sample site may be outside or at a site upwind from the stack emissions. Each situation is unique. Therefore, sample locations must be carefully thought through prior to sampling. For examples of where control locations have served to identify usable background information, see Figures 8.1 through 8.7.[4]

Table 8.2 Listing of Toxic Organics Identified in TO-1, TO-2, and TO-3 as Analyzed by GC/MS

TO-1

Volatile Nonpolar Organics

HALOGENATED HYDROCARBONS
- Bromodichloromethane
- Bromoform
- Bromomethane
- Carbon tetrachloride
- Chloroethane
- *2-Chloroethyl vinyl ether*
- Chloroform
- Chloromethane
- Dibromochloromethane
- 1,2-Dichloroethane
- cis-1,2-Dichloroethylene
- trans-1,3-Dichloroethylene
- Methylene chloride
- Tetrachloroethylene
- 1,1,2-Trichloroethane
- Trichlorofluoromethane
- 1,1,2,2-Tetrachloroethane
- 1,1,1-Trichloroethane
- Trichloroethylene
- Vinyl acetate
- Vinyl chloride

AROMATICS
- Benzene
- Chlorobenzene
- 1,2-Dichlorobenzene
- 1,3-Dichlorobenzene
- 1,4-Dichlorobenzene
- Ethyl benzene
- Toluene
- Styrene

KETONES
- Acetone
- 2-Butanone (MEK)
- 2-Hexanone (MBK)
- 4-Methyl-2-pentanone (MIBK)

OTHERS
- Carbon disulfide

TO-2

All the chemicals which are listed in TO-1 (except *2-Chloroethyl vinyl ether*) with the following additions:

HALOGENATED HYDROCARBONS
- Bromochloromethane
- 1,2-Dibromo-3-chloropropane
- 1,2-Dibromoethane
- Dibromomethane
- Dichlorodifluoromethane
- 1,1-Dichloroethylene
- 1,1-Dichloropropylene
- cis-1,3-Dichloropropylene
- Hexachlorobutadiene
- 1,1,1,2-Tetrachloroethane
- 1,2,3-Trichloropropane

AROMATICS
- Bromobenzene
- n-Butylbenzene
- sec-Butylbenzene
- 2-Chlorotoluene
- 4-Chlorotoluene
- Isopropylbenzene
- Naphthalene
- n-Propylbenzene
- 1,2,3-Trichlorobenzene
- 1,2,4-Trichlorobenzene
- 1,2,3-Trimethylbenzene
- 1,2,5-Trimethylbenzene
- m/p/o-Xylene

TO-3

All chemicals listed in TO-2 with a database search for other peaks identifiable within the NBS.

Table 8.3 Listing of Potential Compounds for Evaluation of NIOSH Indoor Air Quality Methodology

Compound*	ACGIH Acceptable Limits (ppm [mg/m^3])	Source Example
Pentane	600[1770]	natural gas
Hexane	50[176]	rubber cement
Cyclohexane	300[1030]	solvent
Decane	—	copy toner
Benzene	10[32]	paints/gasoline
Toluene	50[188]	paints/gasoline
Xylene	100[434]	paints/gasoline
Limonene	(560)**	lemon odor
Acetone	750[1780]	solvent
2-Butanone (MEK)	200[590]	paints/solvent
Methyl isobutyl ketone	50[205]	resins/solvent
Tetrahydrofuran	200[590]	plastic pipe cleaner
Methyl cellosolve	25[16]	solvent/cleansers
Butyl cellosolve	25[121]	solvent/cleansers
Cellosolve	5[18]	solvent/cleansers
Carbon tetrachloride	5[31] (animal carcinogen)	solvent/cleansers
Tetrachloroethylene	25[170] (animal carcinogen)	solvent/cleaners
1,1,1-Trichloroethane	350[1910]	office partitions
Freon 113	5620	coolant

* Excerpted from Evaluation of Sampling and Analysis Methodology for the Determination of Selected Volatile Organic Compounds in Indoor Air.[3]

** Group consensus of concern level as posed by the World Health Organization

Then, too, where there is a suspect source, bulk sampling may be performed of the material in question and compared with the air sample results for confirmation of the source material. For an example of bulk sample comparisons, see Figure 8.8.[2]

In brief, there is no set strategy for sampling unknown organics. Yet, the approaches mentioned herein have a solid foundation in the experiences of others. Although the cost for GC/MS analyses is high, taking only one sample, in order to avoid the additional cost for analysis of a control sample, may lead to inconclusive or poorly substantiated results. Without an acceptable standard, a control can serve as a gauge, or means for comparison.

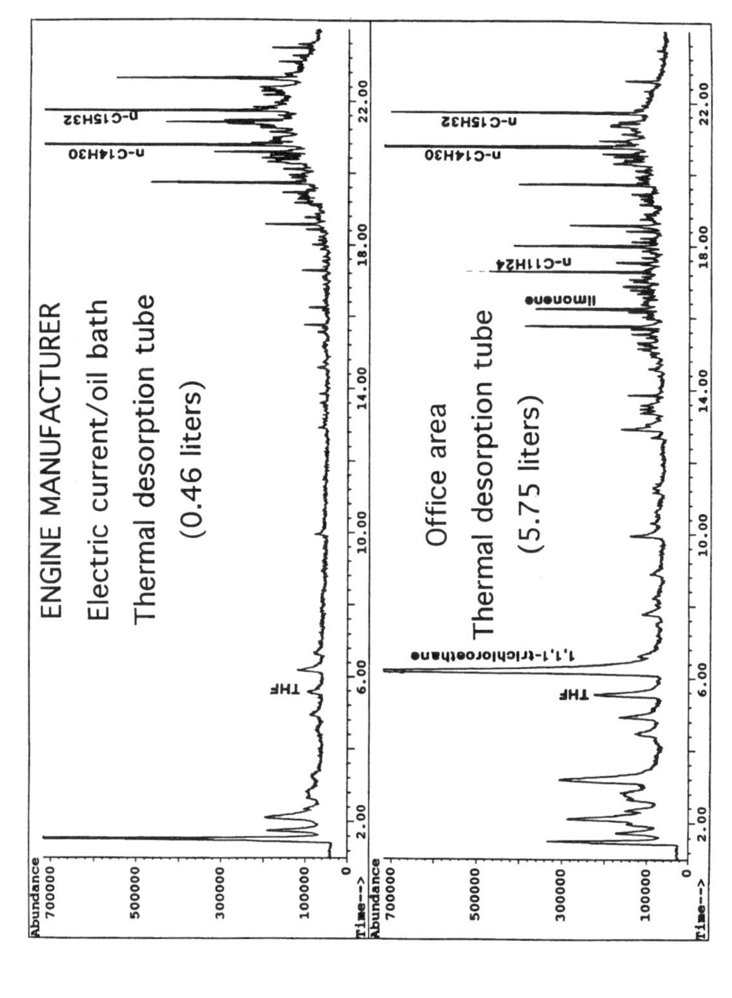

Figure 8.1 Point Source Processing (electric current passing through an oil bath) vs. Adjacent Office Area. There were no organics identified aroung the oil bath process which were not likewise found in the office area. (Courtesy of NIOSH, Cincinnati, OH)

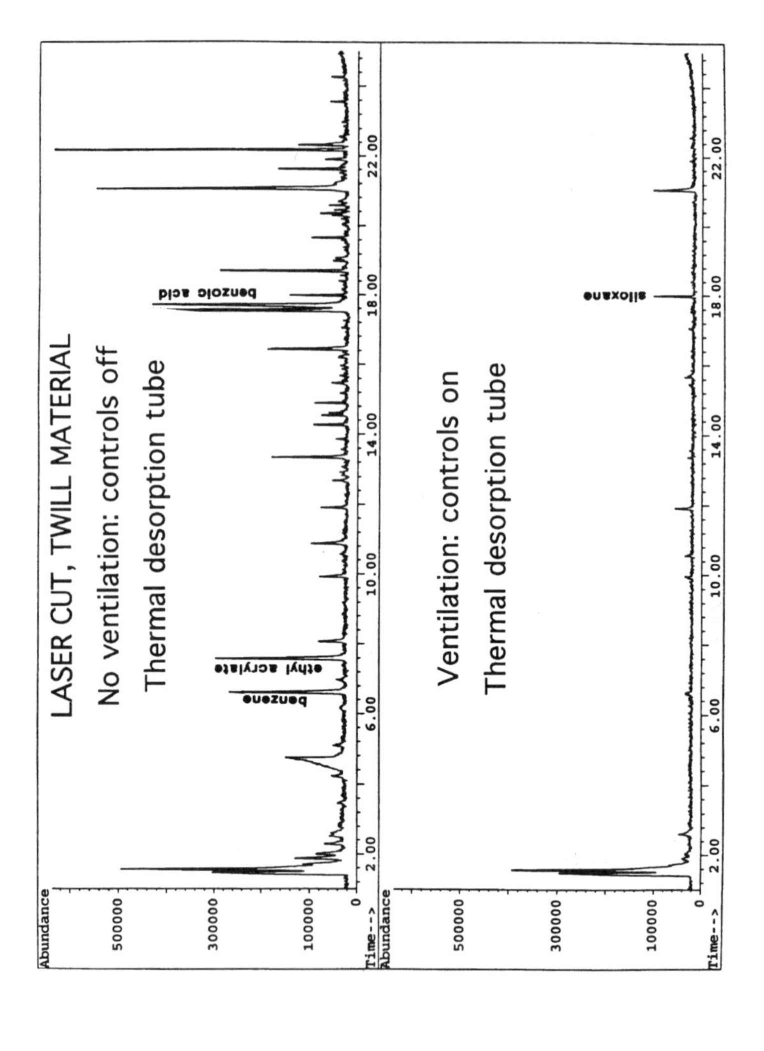

Figure 8.2 No Ventilation vs. Ventilation. Ventilation effectiveness was confirmed. (Courtesy of NIOSH, Cincinnati, OH)

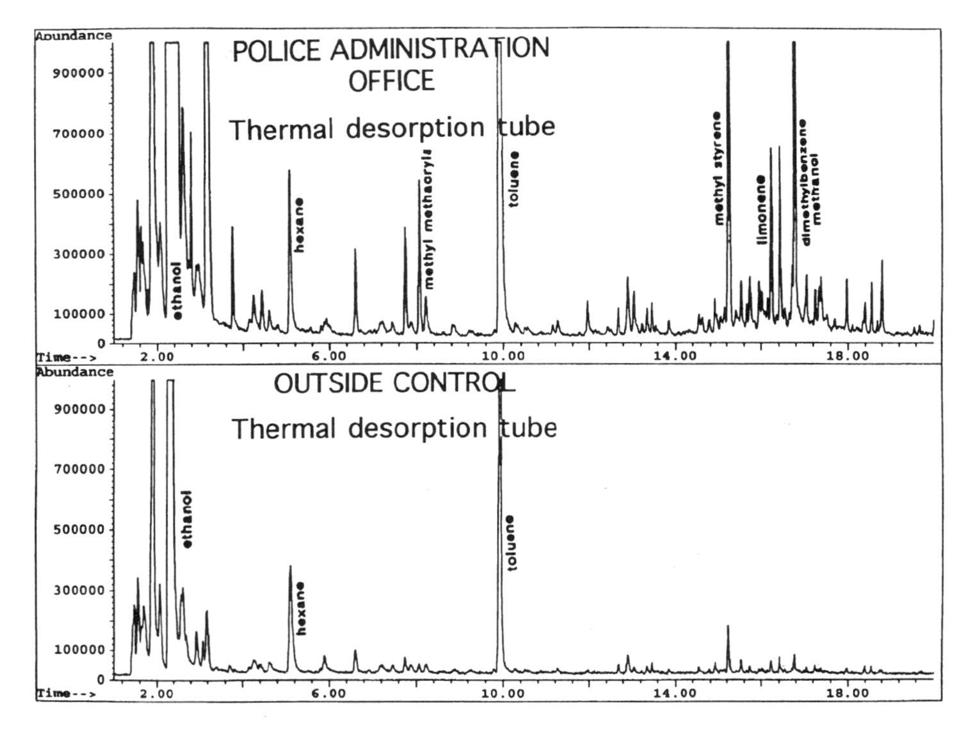

Figure 8.3 Office Space vs. Control (outside area). Control shows many of the same organics found indoors (e.g., toluene) which may otherwise have been mistakenly associated with an indoor source. (Courtesy of NIOSH, Cincinnati, OH)

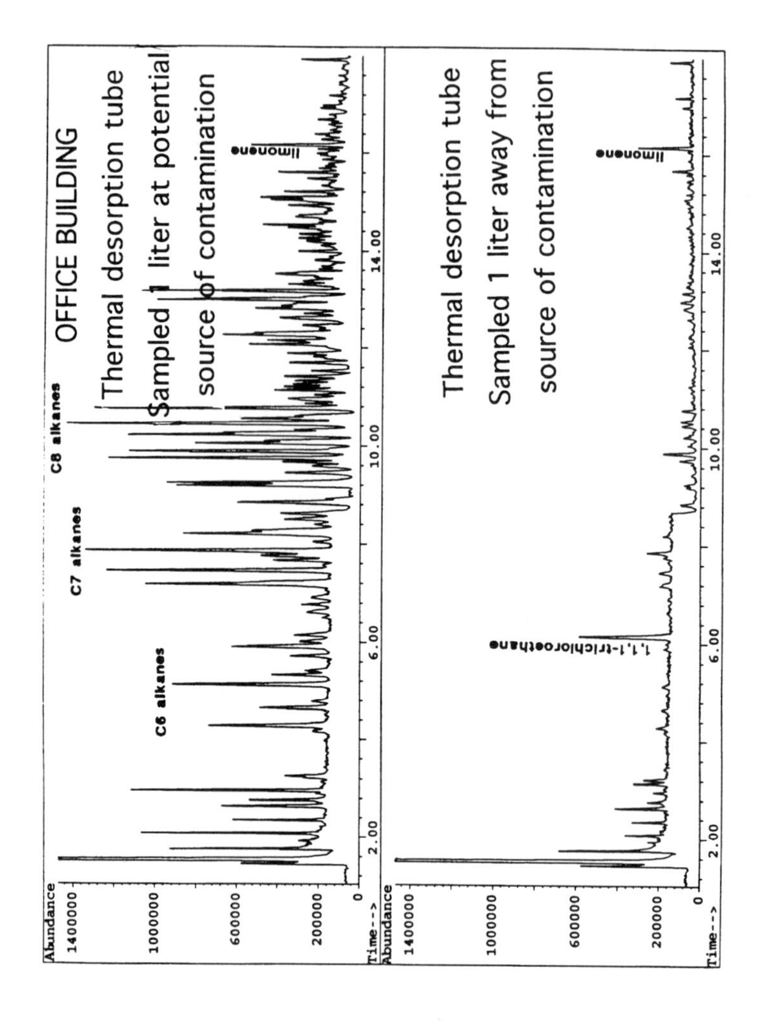

Figure 8.4 Closet Drain in Office Building vs. Control (office area). Speculation was that the origin was petroleum distillates, not gasoline. (Courtesy of NIOSH, Cincinnati, OH)

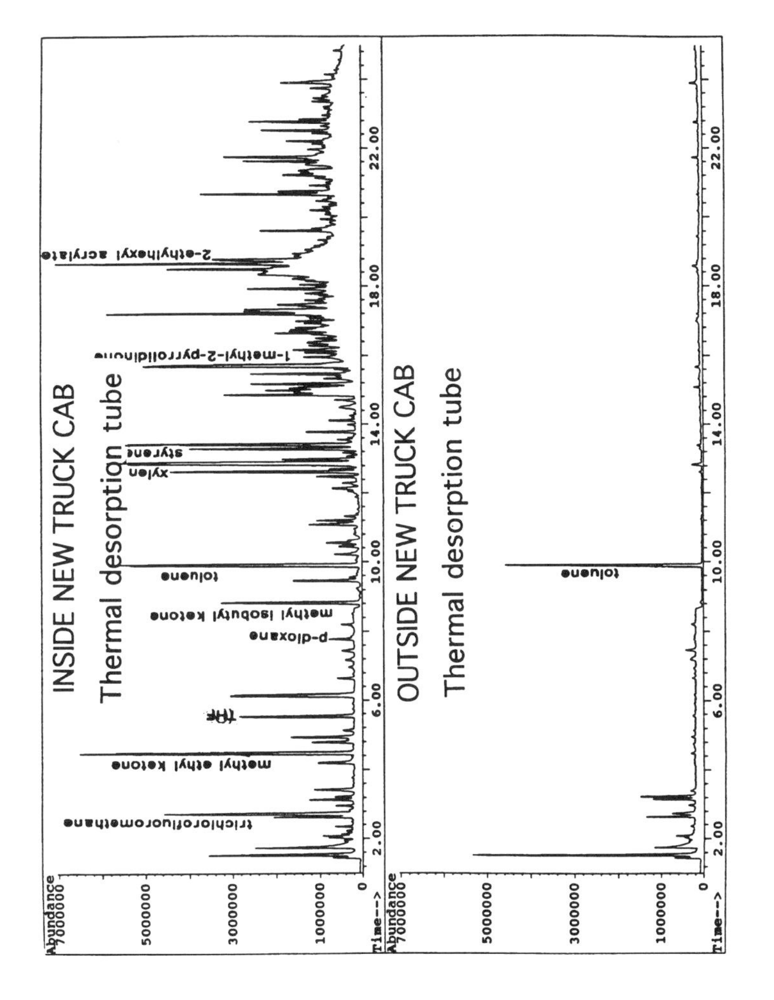

Figure 8.5 Off-gassing of Components in New Truck vs. Control (outside area). The origin of the toluene component was outside. (Courtesy of NIOSH, Cincinnati, OH)

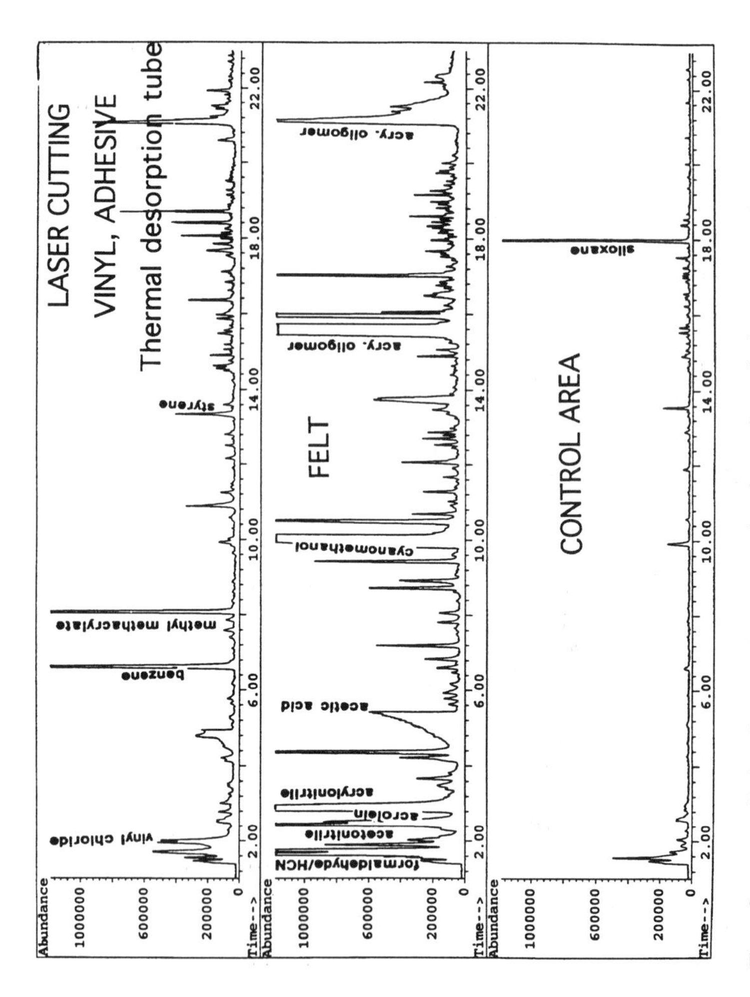

Figure 8.6 Two Separate Point Source Processes (carbon dioxide laser cut of vinyl material/adhesive and carbon dioxide laser cut of felt) vs. Control (away from the cutting activities). [Courtesy of NIOSH, Cincinnati, OH

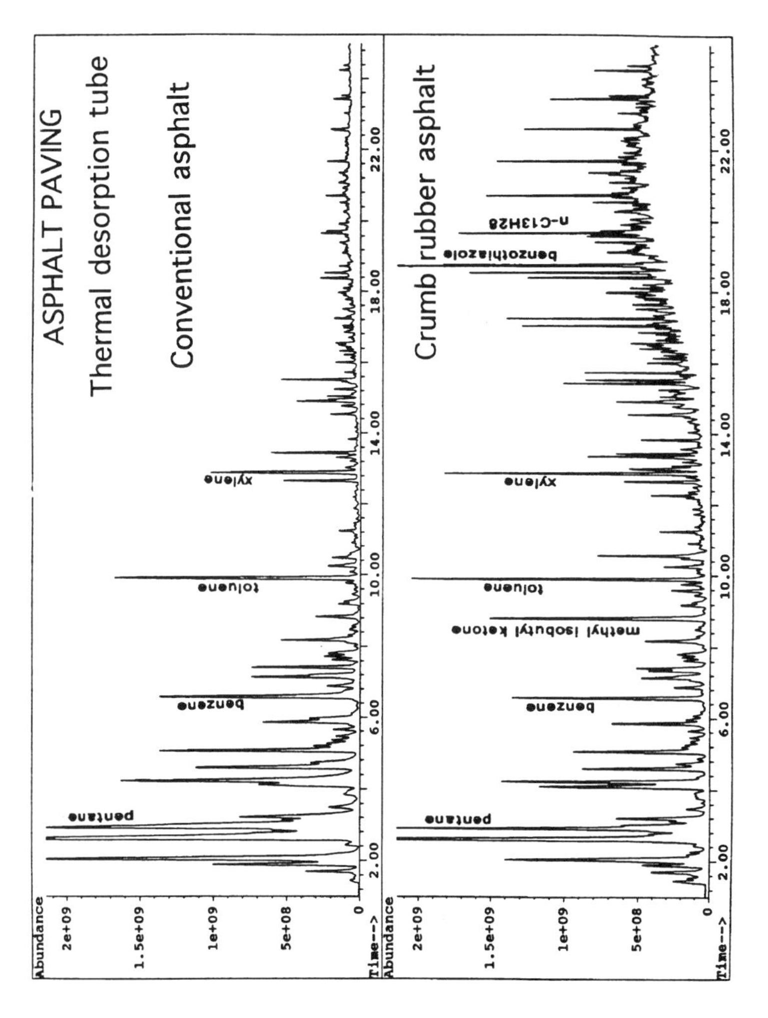

Figure 8.7 Conventional Asphalting vs. Asphalting with Rubber. The asphalt with rubber had methyl isobutyl ketone and benzothiazole. (Courtesy of NIOSH, Cincinnati, OH)

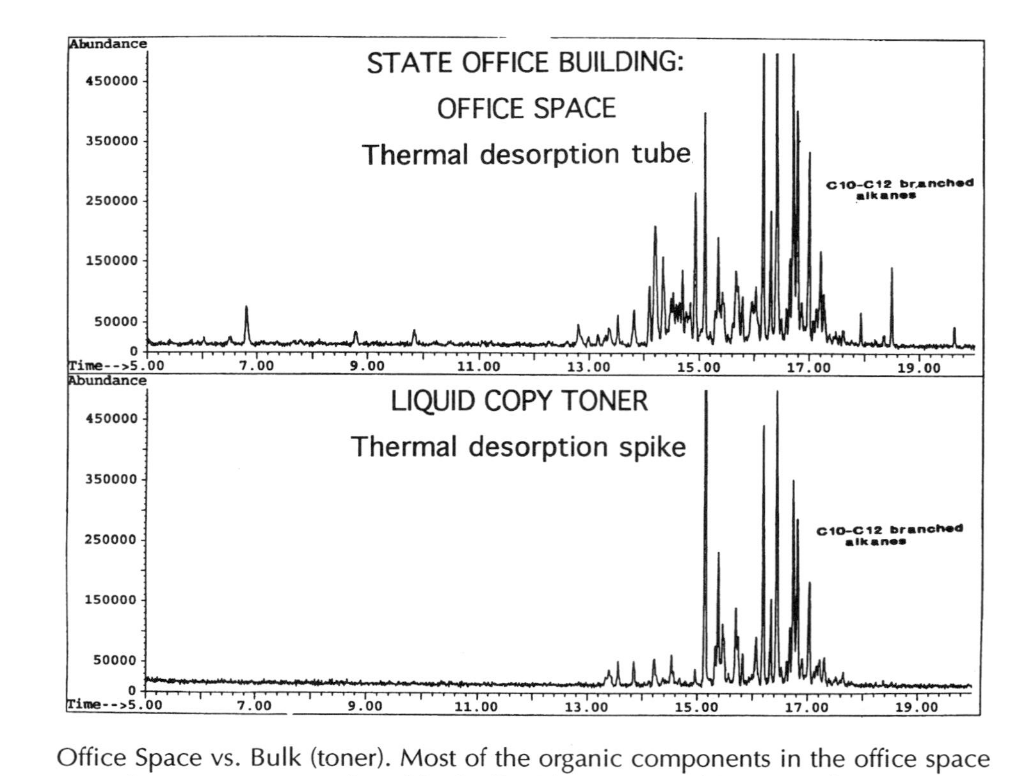

Figure 8.8 Office Space vs. Bulk (toner). Most of the organic components in the office space were due to components found in the liquid copy toner. (Courtesy of NIOSH, Cincinnati, OH)

SCREENING PROCEDURES

Some laboratories and environmental consultants are performing screening for "total organics" prior to deciding on the need for or on the extent of identification indicated. This generally involves, but is not limited to, sampling with charcoal sorbent tubes and quantitation by gas chromatography. This may be performed in conjunction with the collection of samples for identification or prior to making that decision.

The cost of quantitation is one-tenth that of identification. If quantitation can be used as a screening tool, the cost of identification may be circumvented. Facilities managers decide on an action limit for move-in to a building, or they perform routine background sampling in all their buildings. In the latter, when problems arise, a baseline will have already been established for, at a minimum, "total organics."

Although there are no established acceptable limits for the generalized classification of "total organics," some environmental professionals decide on an action limit to serve as a "go-no-go" prior to going to the expense of identification. If using one tenth of the American Conference of Governmental Industrial Hygienists' (ACGIH) individual component standards [as suggested by the American Society of Heating, Refrigeration and Air-conditioning Engineers (ASHRAE)], the lowest limit will be less than 1 mg/m^3. Some state agencies set their own limits (e.g., Texas General Services Commission limit: 0.5 mg/m^3). Based on irritation response levels and a safety factor, one researcher recommends a limit as low as 0.25 mg/m^3.[4] The latter study was performed by the Denmark Institute of Hygiene.

The Denmark study identified irritation levels to organic compounds at 5 mg/m^3. The study involved sixty-two chemically-sensitive subjects, and the total organic level was compared to toluene for analytical purposes. The twenty-two compounds used to challenge the subjects were thought to represent compounds previously found in indoor nonindustrial environments. They are listed in Table 8.4.

Another researcher recommends a limit of 0.30 mg/m^3 with no more than 0.06 mg/m^3 for any one component.[5] If the potential exists for the presence of an extremely toxic substance, the limit may even be lower (e.g., 10 percent of the ACGIH standard for acrolein is 0.01 mg/m^3). These issues are primarily those of indoor air quality concerns.

Then, too, semivolatiles may be present, contribute to the total organic compound composition of the air. Total organic quantitation prior to identification may serve as a means for determining the need to have the semivolatiles identified along with the total volatile organics. Where thermal desorption analysis is performed, the sample can be lost on the volatile analysis. (Some sorbent systems capture semivolatiles.) Semivolatiles are not analyzed by the same protocols as volatiles. A separate analysis must be performed.

Table 8.4 Compounds Used to Challenge Subjects in the Denmark Study[6]

n-Hexane
n-Nonane
n-Decane
n-Undecane
1-Octane
1-Decene
Cyclohexane
3-Xylene
Ethylbenzene
1,2,4-Trimethylbenzene
n-Propylbenzene
α-Pinene
n-Pentanal
n-Hexanal
Iso-propanol
n-Butanol
2-Butanone
3-Methyl-3-butanone
4-Methyl-2-pentanone
n-Butylacetate
Ethoxyethylacetate
1,2-Dichloroethane

If solvent desorption of a sorbent (or, in some cases, bulk air sampling) is anticipated, the same sample taken for identification purposes may be used for screening. The end result is a decision as to whether volatiles alone are to be identified or the semivolatiles are to be processed as well. Volatiles and semivolatiles often require two separate identification scans and are better identified by treating them as two separate samples. If properly anticipated, the appropriate number of samples and/or sorbents can be planned and a return visit avoided.

Sampling Methodologies for Screening

The most commonly used approach to screening for total organics is the use of charcoal sorbent tubes. Charcoal sorbents are a good general purpose sampling media, and they are capable of capturing both nonpolar volatile and semivolatile organics. The other sorbents are less inclusive and more targeted toward specific groups of compounds. Keep in mind, a very few of the polar organics are captured, but most (e.g., aliphatic and aromatic amines) pass through the sorbent without adsorbing to charcoal sorbents.

Bulk air sampling (e.g., samples collected by bag or Summa® canister) is another means that may be used to collect the sample. In these methodologies, the air is captured while the analytical method limits the material included in a total organic screening. One problem with this method which would otherwise appear ideal is laboratory limitations.

Some commercial laboratories are equipped to quantitate air samples collected in a bag or evacuated container. Others are equipped to handle direct injection of the air sample into the analytical equipment. This technique may be particularly useful where large volumes of air have been collected (e.g., greater than 5 liters), and the same sample is intended for both screening and identification. The biggest drawback to this approach is the limited number of laboratories capable of performing the combined analyses, whereas the charcoal tube approach is readily performed by most commercial laboratories. Where some laboratories may be capable of the screening process, they may not be able to perform the identification. However, a laboratory is generally able to perform bulk air sample identification only.

Another consideration in using bulk air sampling is the limit of detection. Whereas a 20 liter sample volume on a sorbent may permit detection down to 0.05 mg/m^3, a bulk air sample will only detect down to 1 mg/m^3 or, in certain special cases, to 0.5 mg/m^3.[7] If the screening process involves a lower limit of 0.25 mg/m^3, the bulk sampling approach is not feasible.

If charcoal tubes are used, the sampling pump flow rate should be below 200 ml/min. The greater the air volume, the greater the detection of limit. Yet, when considering air volume requirements, the environmental professional should exercise caution!

Large sample volumes may result in vapor breakthrough on the sorbent which will invalidate the results. A 10 liter sample, taken at 200 ml/min., will require 50 minutes of sampling time. In most cases, this should suffice in detecting levels as low as 0.1 mg/m^3 where a 20 liter sample permits detection down to 0.05 mg/m^3.

More details on sampling methodologies and volatile organic identification are discussed later within this chapter. Most sampling methods used for identification may also be used for screening. Thus, samples may be taken simultaneously in situations wherein the requirement for identification is anticipated. Screening methodologies (e.g., evacuated containers) do not apply conversely!

Analytical Methodologies for Screening

The screening procedures are limited mostly by the analytical methodology used for "total organics." This includes the retrieval of the captured organic components (or desorption), the reference standards, and the analytical equipment.

Where a charcoal sorbent has been used for capture, analyte extraction is generally performed with the solvent carbon disulfide. Carbon disulfide may or may not desorb all the organics, and if it does desorb them, it may interfere with the analysis. It is the most commonly used desorbing solvent due to its overall qualities and use for desorbing/analyzing most organics. At the same time, desorption with carbon disulfide will mask or not be able to desorb some organics. Yet, this is the case with any chosen desorbing chemical, and carbon disulfide is generally chosen due to its least restrictive qualities.

The laboratory reference standard generally used to determine quantity of total organics is hexane. Laboratory preferences vary, and sometimes the standard used is a higher molecular weight alkane. Some laboratories use methane to determine total organics in compressed air delivery systems. Others use decane or higher. Whichever alkane is chosen, one chemical standard does not compare well with other compounds. True quantitation can only be obtained by comparing each component compound with its own. Quantitation of total organics only provides a ball-park number.

Gas chromatography is used for quantitation of the desorbed chemicals or of the bulk air sample. Where a sorbent sample is being analyzed, an aliquot of the desorbing solvent and desorbed material is injected into the instrument. This is, in turn, carried by an inert gas stream through a column which separates the components. This separation is based on the boiling point of the individual components and on their affinity for the packed material or coating on the column. Then, as they emerge from the column at different rates, the separated components are passed through a detector. In most cases, the flame ionization detector is used.

The flame ionization detector will detect almost all organics. Yet, its response is greatest to hydrocarbons, decreasing with increased substitution of other elements (e.g., oxygen, sulfur, and chlorine). As the complexity increases, there is a decrease in detection sensitivity. For this reason, most organics will be detected, but quantitation response is diminished with increased complexity. Other detectors which may be used are presented in Table 8.5.

Although additional sampling and/or analyses may ultimately be indicated, screening may save time and/or money. Where the total volatile organic compounds is less than 0.25 mg/m^3, some investigators may opt to limit further analyses. Where semivolatiles appear predominantly in the analytical process, sampling and analytical procedures may be adapted to accommodate the semivolatiles as well as the volatiles.

Table 8.5 Types of Detectors and Detection Capabilities

Type of Detector	Detection	Poor/No Detection
Flame ionization detector	aliphatics aromatics some organics with oxygen, sulfur, and halogen molecules	inert gases inorganics
Photoionization detector	higher molecular weight aliphatics aromatics halogenated hydrocarbons some inorganics arsine, phosphine, and hydrogen sulfide	methane
Electron capture detector	organics with halogens, cyano and nitro compounds	aliphatics aromatics alcohols ketones

Excerpted from *Air Monitoring for Toxic Exposures.*[8]

SOLID SORBENT SAMPLING

Solid sorbents are any of a number of solid materials which can capture, or adsorb, specific compounds that can later be extracted, or desorbed, from the collection medium. There is no one single sorbent which can be used to sample for "all organics." Herein lies the greatest dilemma to sorbent sampling. All organics are not captured and, if captured, may not be retrievable.

Some desorbing chemicals completely drive the compound(s) from the sorbent and will not interfere with analysis while others can be desorbed only partially. Some captured components may require special processing to be displaced, but their presence must be suspect before the analyst will use a different desorbing material. Then, special processing, required for select compounds, generally interferes with the analysis of the other unknowns. In brief, retrieval of all captured material is rarely feasible.

On the more positive side, solid sorbents concentrate large sample volumes onto one tube. Where a chemical is present in extremely low quantities, it may not be detected without concentrating the sample onto a sorbent. Sample volumes available for analysis may exceed 250 liters for thermal desorbent tubes as compared with the greatest volume available through bulk air samples which is 1 liter.

With thermal desorption, total sample air volume which is actually analyzed may be less 2 liters (as per NIOSH procedures) or as high as 250 liters (which is the maximum air sample volume for EPA Method TO-1). However, bulk air samples use 0.5 to 1 liter of air sample when managed cryogenically.

On the other hand, chemical desorption of the sorbent has different features. Typically, two one-thousandth of a chemically desorbed sample is analyzed. So, a 10 liter air sample provides the equivalent of 0.01 liter of air sample. In order to provide sensitivity similar to that of cryogenically treated bulk samples, a volume of 500 liters would be required. Yet, several sample runs are feasible with chemical desorption where only one pass is possible for one given thermal desorption sampling tube.

The most sensitive methodology for unknown organic sampling are the use of solid sorbent tubes with thermal desorption. If, however, both volatiles and semivolatiles are sought or further analyses are anticipated, additional samples may be collected.

Solid Sorbent Characteristics

Although the published methodologies specify the type of sorbent to use, an understanding of special uses and limitations of sorbents will assist in developing strategies and deciding on the approach which is most applicable to a given situation. Sorbent capture efficiency is affected by environmental conditions and sampling parameters.

Sorbents collection efficiency is impacted by temperature and humidity. Increased temperatures result in decreased adsorption, irrespective of the type of sorbent used. High humidity (greater than 80 percent) results in decreased adsorption by certain sorbents for some analytes. Those sorbents which are most likely to be impacted are charcoal, silica gel, carbon molecular sieve, and alumina. See Table 8.6 for additional information.

Sorbent capture efficiencies are based on known sampling parameters. These include the following:[9]

Flow rate—Increased flow rates result in decreased adsorption efficiencies. A moderate flow rate for most organics is at or less than 200 milliliters per minute.

Concentration—Elevated concentrations of competing organics will increase the sorbent loading, resulting in breakthrough. Breakthrough is the passage of a chemical or chemicals through the sorbent without being captured or after capture being displaced by another chemical for which the sorbent has a greater affinity.

Table 8.6 Characteristics of Solid Sorbent Media for Sampling Organics Compounds

Type	Chemical Preferences	Affected by High Humidity
Charcoal	Nonpolar volatile and semivolatile organics	Yes
Silica gel	Polar volatile and semivolatile organics	Yes
Carbon molecular sieve (e.g., Carbosieve)	Nonpolar highly volatile organics	Yes
Tenax porous polymer (e.g., Tenax GC)	Nonpolar semivolatile organics	Low Impact
Porous polymer (e.g., Chromosorb 102)	Selective for specific components (dependent on type of polymer used)	No
Alumina gel	Polar high molecular weight organics	Yes
Florisil	Polychlorinated biphenyls and other pesticides	No

Highly volatile organics have a boiling point between 50° and 100°C.
Volatile organics have a boiling point between 50° and 260°C.
Semivolatile organics have a boiling point between 240° and 400°C.

Sample volume—While increasing the analytical efficiency, high air volumes will also increase the chances of sorbent loading. The amount of air sampled should be based on the concentration anticipated, the limits of the sorbent, and the limits of the analytical methodology.

Competition between chemicals—Chemicals have a differing affinity in the different adsorbents. Those with a stronger affinity will displace those with a lesser attraction.

Sorbent particle size—The smaller the adsorbing particles the greater the surface area, therefore there is an increase in adsorption. Some of the sorbent surface areas are listed in Table 8.7.

Amount of sorbent—An increase in sorbent volume will result in an increase in capture material. Sorbent tubes contain between 100 and 1000 milligrams of material.

Type of sorbent—The solid sorbents vary not only in their ability to capture specific compounds but in their feasibility for extraction within the laboratory. Many toxic organic compounds have been tested with the various sorbents for capture and extraction efficiency, but many have not. Particularly difficult are those compounds which are not listed as toxic or that are irritants only. These are likely to have been excluded from any research or development studies.

In a study performed by NIOSH, two separate sorbents were used to sample within the same time frame at the same sample site (e.g., a rubber molding facility). One was sampled by thermal desorption tube with a carbon based sorbent and analyzed by GC/MS. The other was sampled by charcoal tube, chemically desorbed, and analyzed by GC/MS. The results, represented in Figure 8.9, favored the thermal desorption sampling and analytical approach over the other.[2] Compounds missed in the chemically desorbed charcoal tube included aliphatic amines, sulfur dioxide, and carbon disulfide (which was the desorption compound used for most chemical extractions from charcoal tubes). The conclusion is that thermal desorption is more efficient than chemical desorption methodologies.

In another study, NIOSH evaluated adsorption on multibed thermal desorption tubes as represented in Figure 8.10.[8] These were, in order of occurrence (e.g., direction of sample flow), Carbopack Y, Carbopack B, and Carboxen. The first captured the least volatile compounds. The second captured more volatile compounds, and the third captured those with the highest volatility. Some layers collected portions of that which was collected on an adjacent layer. Had only one sorbent been used, the other compounds would have been lost, not identified.

Table 8.7 Sorbent Surface Areas

Sorbent (Source)	Surface Area (m^2/gm)
Carbotrap C (Supelco)	10.0
Carbopack C (Supelco)	12.0
Tenax GC (Enka)	23.5
Tenax TA (Enka)	35.0
Carbopack B (Supelco)	100.0
Carbotrap (Supelco)	100.0
XAD-2 (Rohm & Haas: Amberlite)	364.0
Carboxen-564 (Supelco)	400.0
Carboxen-569 (Supelco)	485.0
Carboxen-563 (Supelco)	510.0
Carbosieve SIII (Supelco)	800.0

Excerpted from *Evaluation of Sampling and Analysis Methodology for the Determination of Selected Volatile Organic Compounds in Indoor Air.*[10]

Decide on the sample site location, then decide how the sample should be collected. This involves sorbent selection and a determination as to the desired air flow rate and sample volume.

Sorbent types have previously been discussed. Selection of type may be limited by one of the EPA or NIOSH published methodologies, or the laboratory may have alternate approaches. The choice will be based upon information sought and the capabilities of the laboratory.

Some of the more important parameters have been provided should the environmental professional and laboratory choose to delve into this area more creatively. Should this be the case, however, adsorption/desorption efficiencies will require more intensive background studies. Many laboratories have already gone through this process, but most have not. When delving into unpublished approaches, experience and a track record are desirable. In choosing a commercial laboratory, proceed with caution.

The air flow of the sampling pump should be calibrated according to published methodologies and established at a predetermined flow rate which has either been published or dictated by the servicing laboratory. The rates are, however, rarely singular. The environmental professional is given a range from which to choose or a maximum flow rate. The high range should not be exceeded at any cost. Rates higher than that which is recommended will increase the possibilities for breakthrough on the sorbent. The lower end of the range is less vital unless the low rate approaches diffusion uptake rates

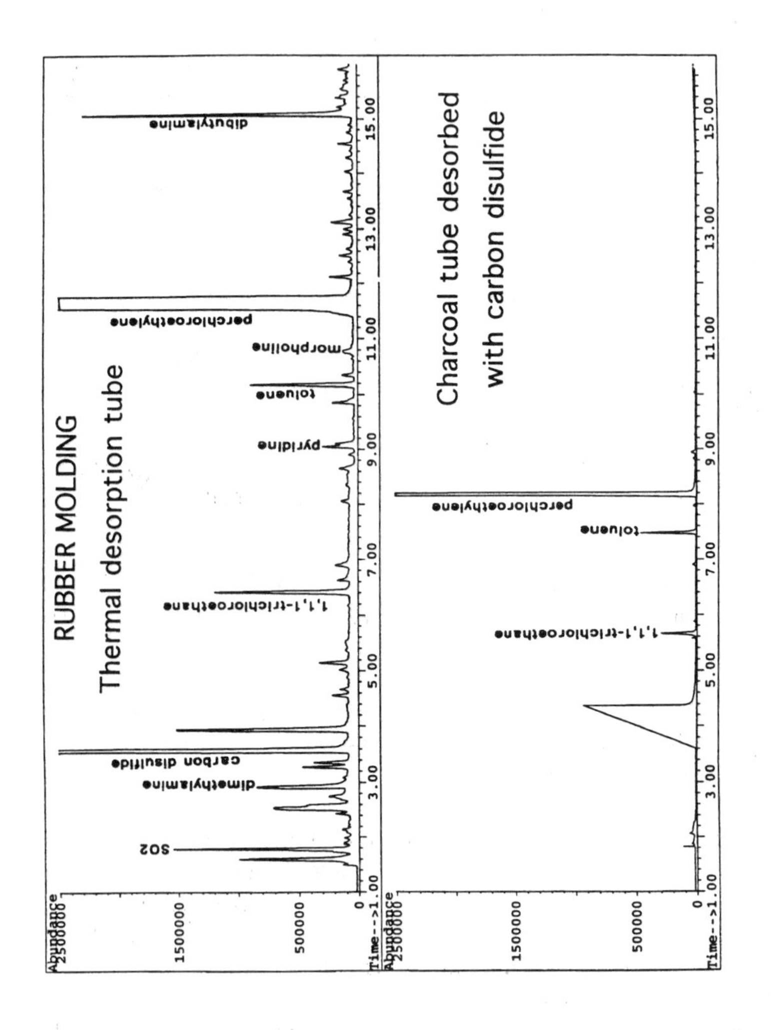

Figure 8.9 Thermal Desorption vs. Chemical Desorption (Courtesy of NIOSH, Cincinnati, OH)

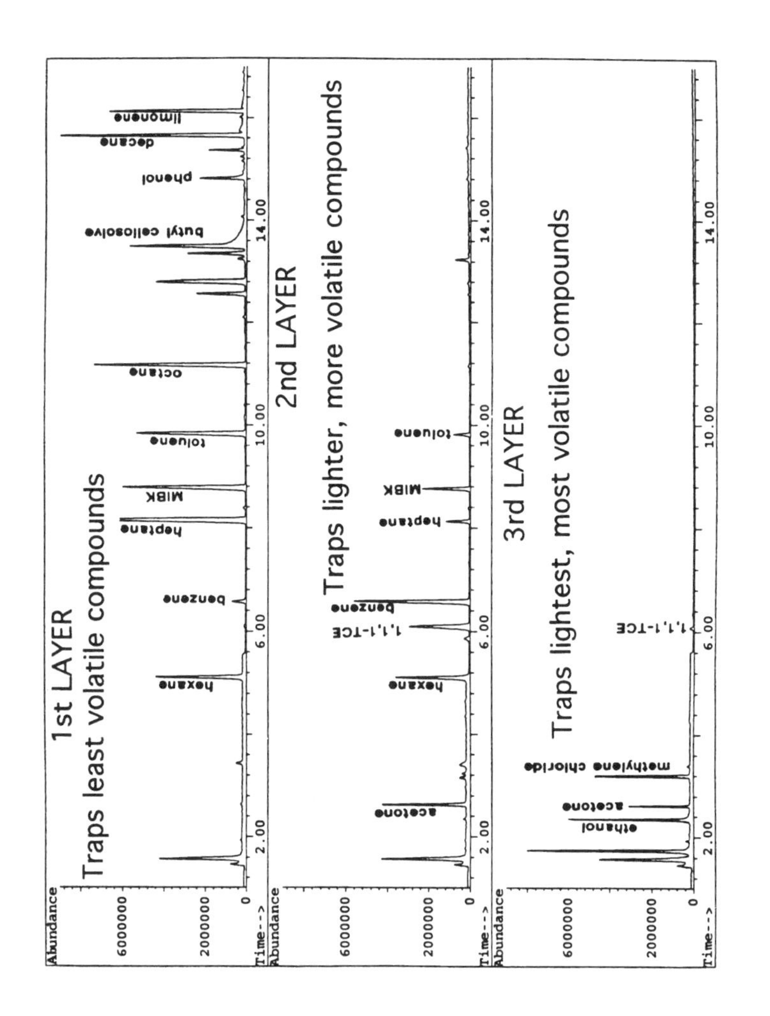

Figure 8.10 Multibed Thermal Desorption (Courtesy of NIOSH, Cincinnati, OH)

(i.e., 5 cubic centimeters per minute). All choices in between are discretionary. Relevant issues, which may be used to adjust these choices, include the following:

> **Desired sample duration**—If exposures are anticipated within an abbreviated time period, that window of occurrence will provide information regarding a worst case scenario. Where exposures are thought to be consistent throughout the day, sampling may be performed within that time period. Where exposures are thought to be consistent but a quick sample is desired, the duration may be reduced accordingly.
>
> **Desired air volume**—Higher anticipated sample volumes will require adapting the flow rate, along with the sample duration, to assure a sufficient amount of air has been sampled. For short duration samples, requiring large volumes of sampled air, the air flow rate will need to be at the top end of the range. For longer durations, requiring small air volumes, the low end of the range is indicated.

Desired air volume is based upon a combination of the sampling methodology limits and on the anticipated concentration of organics in the air. The sampling methodology limits are published or provided by the laboratory. The anticipated concentration of organics can, however, be a crap shoot. Unless there are visible emissions, odors that are evident to all passers-by, or the point source is known, concentrations are likely to be low. If monitoring an industrial process, stack emissions, or hazardous waste emissions, concentrations are likely to be high. Professional judgment is necessary.

Once the parameters have been decided, set the pump(s), collect the desired air volume, and send the sample(s) to the laboratory of choice, along with the sampling data. After recording the sample location, pertinent information which should accompany each sample to the laboratory includes the following:

- Sample name or number
- Air volume sampled
- Temperature where the sample was collected
- Humidity where the sample was collected
- Barometric pressure

Care should be taken so as not to allow contamination of the samples during shipping from outside sources [e.g., a bulk organic sample shipped with the sorbent tube(s)]. Samples may be damaged in shipping, and one may not suspect a problem until the analysis has been completed and questionable results are provided to the originator. Sampling is not complete until samples have safely reached the laboratory and been analyzed.

Analytical Methodology

At the laboratory, the sample is desorbed either thermally or with a solvent. Solvent desorption involves extraction of the sampled material with a desorbing substance. One hundred milligrams of charcoal are normally extracted with 1 ml of carbon disulfide. Then, an aliquot of 1 μl of the desorbent and desorbed material is injected for analysis. In other words, only one thousandth of the material actually sampled is analyzed.

Thermal desorption, on the other hand, involves the installation of the sample tube directly in line with the analytical equipment. The tube is heated to a predetermined temperature, and many of the captured components are thermally driven from the sorbent. It is important that the environmental professional coordinate with the laboratory prior to sampling in order to obtain information regarding the laboratory's equipment and the type of tube to purchase. In many cases, the laboratory will prefer sending one of their own prepared and thermally cleaned tubes.

Identification analyses are performed by gas chromatography/mass spectrometry, referred to as a GC/MS. This procedure combines the quantitative and chemical separation features of the gas chromatograph with chemical unknown identification. The mass spectrometer takes the chemical components which have been separated by the gas chromatograph, fragments each separate unknown, and records the fragmentation pattern. This pattern is, in turn, compared with features of known chemicals. The comparison is typically made by a computerized library search of other patterns. Then, the computer provides information as to the best fit for each of the components.

Rarely is the match a hundred percent, while ninety-five percent is considered a very good fit. The chemist then decides if the match is significant for identification or whether the chemical can not be identified through the library. Most GC/MS libraries have over 70,000 on file. Some have as many as 150,000 chemicals. These should identified most of the more common or widely encountered chemicals.

BULK AIR SAMPLING

A slightly less sensitive yet more comprehensive sampling approach is that of bulk air sampling. Most bulk air samples represent the average of "all chemical components" over the sample duration. Bulk air samples are neither exclusionary in scope nor impacted by the capture and desorption parameters that limit sorbent sampling. One collection sample may also be used for more than one analysis which is true of chemically desorbed sorbents as well. Thermal desorption of sorbents is typically a one shot affair, depending on the thermal desorption equipment used.

Bulk air sampling allows for some degree of sensitivity. It is comprehensive in scope of chemical components captured, and it permits more than one chance at analysis.

Sampling Methodologies

There are several means for performing bulk air sampling. Not all are equally competent. Each method varies in equipment accessibility, ease of handling, and laboratory capabilities.

Evacuated Samplers (*e.g., Summa® Canisters*)

Although evacuated samplers are simple and easy to use, they are expensive and always require special laboratory preparation. Then, too, laboratories vary in their ability to handle the different types of samplers. In order of simplicity, the three main samplers are vacuum cans, ambient air tubes, and ambient air canisters.

Vacuum cans are evacuated cans which are capable of collecting relatively small volumes of air (e.g., less than 400 milliliters). They look like shaving creme canisters. They are easy to use and easy to transport. The specialty sampling ridged cans are evacuated by any of a number of manufacturers (e.g., MDA Scientific) and typically back filled with enough nitrogen to prevent the cans from collapsing, while diluting the sample by as much as half. Their shelf life is up to one year.

To use an evacuated can, simply expose and depress the actuator button for a designated time period (e.g., 10 seconds), record the pertinent information, secure the opening, and ship to the laboratory. After recording the sample location, pertinent information which should accompany each sample canister to the laboratory includes the following:

- Sample name or number
- Temperature where the sample was collected
- Humidity where the sample was collected
- Barometric pressure

As containers have been known to explode, caution should be taken where temperatures are elevated both in the sampler storage, sample collection, and shipping. Some laboratories can analyze down to 1 mg/m^3, as compared with a methane standard. The typical detection limit is 2 mg/m^3.

Ambient air samplers are an adaptation of samplers used for sampling compressed air quality. They typically look like a large, sealed test tube. Although intended for use in-line with a compressor, these samplers have been adapted for sampling ambient air quality. Their volume capacity is typically less than that of the cans (e.g., less than 50 ml.). These samplers must be prepared by the laboratory which will perform the analysis, and their shelf life is limited (e.g., less than one week). Some laboratories can analyze down to 0.5 mg/m^3 total organics, as compared with a methane standard. The typical detection limit is 1 mg/m^3.

To use an ambient air sampler, simply insert the special rubber, squeeze bulb prongs into the tube receptacle. Force air into the tube by squeezing the bulb a predesignated number of times (e.g., three times). Withdraw the squeeze bulb, and send for analysis along with the pertinent information listed above.

Ambient air canisters are specialty samplers. One such sampler is referred to as the Summa® canister. It comes in various sizes (ranging from 0.85 to 33 liters in volume). The preferred, easiest to obtain size is 6 liters. Each canister has been treated with chrome-nickel oxide internally to prevent rusting and minimize organic adherence to the surface of the container, and the canister can be used in its evacuated stage or with a sampling pump. Prior to use, each canister must be cleaned and prepared by a laboratory which has canister analytical capabilities.

The canister preparatory process takes up to 24 hours per can, so the laboratory will require some lead time. As part of the planning process, the opening may also be fitted with a flow control device or the inlet remain as is. It depends on the sample duration desired by the environmental professional. For a 6 liter canister, the valve provided with the canister may be opened, and the sample collection time will take less than 30 seconds. For a longer sample duration, the opening may be fitted with special control valve or critical orifice with a known, calibrated flow rate. These devices will permit the vacuum within the canister to draw the sample air over a time period of up to 24 hours.

To use an evacuated canister after it has been cleaned and fitted with special flow control (when required), the inlet or control valve is opened. The ambient air is drawn into the canister by vacuum. Start and finish times should be recorded along with pertinent sample information and location. If samples are taken over an extended period of time, the temperature and humidity should be checked routinely and averaged for reporting purposes. The sample locations may be remote to the site of the canister by using air tight seals, extension tubing, and allowing the ambient air to fill the extension tube prior to sample collection. Care should also be taken so as to choose tubing which is inert and will not off-gas organics. Canister detection is the same as that of an ambient air samplers without special cryogenic processing. Cryogenic processing allows for detection down to 5 ppb (or 0.005 mg/m^3), as compared to methane. This number varies, depending upon the laboratory standard or standards used and the types of components to be identified.

Ambient Air Sampling Bags

Air sample collection bags are specialty sampling equipment and are constructed of any of a number of synthetic materials (e.g., Tedlar and Teflon). Care should be taken to confirm that the bag material chosen for sampling will not off-gas materials which the environmental professional may be attempting to capture, components that may interfere with or may be lose through surface adhesion to the sample bag. Commercially available bags range in holding capacity from 0.5 to 120 liters, and sampling does require a pump.

In order to perform ambient air sampling, a special bag sampling system must be developed or purchased. This involves retrofitting an airtight container (which is larger than the fully-filled bag) with attachments to allow access of the bag valve to the ambient air. Install a sampling pump and tube into the container. Assure all materials used in the construction of the system are not chemically reactive, and test for air tightness. The same system is reusable. See Appendix 7 a commercially available "integrated bag sampler."

A chemically nonreactive, sample bag is installed and sealed within an airtight container which is larger than the fully-inflated sampling bag and has chemically nonreactive, valve-fitted inlet and outlet ports. Where the inlet is connected to the sample bag, the outlet is connected to a vacuum pump. As the vacuum pump draws a vacuum inside the sealed container, the sample bag draws in ambient air to replace the void which is created inside the sealed container. In this manner, the collected air does not pass through a previously-contaminated sampling pump, and the limited contact with various surfaces (to which chemical may adhere) that might result the loss of chemicals from the air being sampled.

The sample duration may be controlled by the flow rate of the sampling pump, and upon completion the bag valve stem is closed. The container seal is released, and the sample is removed/prepared for shipping. Be sure to record all pertinent information as discussed previously (e.g., sample name or number, temperature, humidity, and barometric pressure).

Shipping poses several problems which are unique to this method. Due to the failed rigidity of the sampling bag, an extreme change in atmospheric pressure (which does occur in air transportation) may result in an expansion of the bag to the point where its integrity is compromised if the bag is full. It is not unusual for sample bags to arrive at their destination with nothing inside. The way to avoid this problem is to collect only half the capacity of the bag, or ship the bag in a rigid, pressurized container.

Detection is the same as that of an ambient air tube (e.g., as low as 0.5 mg/m^3 as compared with a methane standard). Analysis is also performed in a similar fashion.

Analytical Methodologies for Bulk Air Sampling

All sampling methods mentioned may be analyzed by direct injection of the air sample into a GC or GC/MS. Not all GC nor all GC/MS equipment is outfitted to accommodate direct injection of ambient air samples, and rarely is a GC/MS fitted with special cryogenic concentration of samples.

The amount of actual sample which is injected into the analytical equipment is typically 1 milliliter. In the case of ambient sample tubes and bags, the sample is undiluted. Where an evacuated can has been used, however, the sample is generally diluted in half by the preexisting nitrogen content which was part of the can preparatory process. This means the can samples which arrive at the laboratory are not as concentrated as the tube and bag samples, resulting in a reduced detection with the cans. The canister sample may be analyzed in a fashion similar to that of the tubes and bags, but it is much preferred to cryogenically processing the canister contents. Although the EPA "TO" Method 12 involves the use of an ambient air canister and analysis by cryogenic treatment and injection into the GC/MS, commercial laboratories generally do not provide this service. All bulk air sampling methods may be used for screening of "total organics" by GC or for identification of suspected large quantities of organic components by GC/MS if the laboratory is outfitted to perform the service.

Cryogenic processing of bulk air samples is typically associated with the GC/MS. It is not generally used for screening of "total organics." The cryogenic process takes 100 to 1,000 milliliters of sample and concentrates it prior to sample release for analytical processing. As compared with the other direct injection samples, this method secures one hundred to one thousand more sample than the other bulk air samples with "most all of the original ambient air sampled." This is only surpassed by thermal desorption of sample volumes in excess of 1 liter, and the sorbents are highly selective in their ability to deliver a comprehensive sample.

It is worthy of note that even if all the original ambient air is present, humidity impacts the analysis, particularly where a sample is to receive cryogenic process. Not only are the chemical components concentrated by the process, but the water is concentrated as well. Although water can be filtered from the system, polar compounds have a tendency to be attracted to the water and subsequently get filtered out with the water. They may not be identified. Should the water be allowed to enter the system, it tends to plug the system causing other problems. The lower the humidity, the better the analysis. Where high humidity levels are normal, some areas of the country are poor locations for considering reliable canister sampling.

At the present, there is an EPA list of ambient air reference standards for cryogenically treated canister samples. These are listed in Table 8.7 and have come from the EPA Method TO-14.

Table 8.7 List of Chemical Reference Standards in EPA Method TO-14

Chemical Name (in order of retention times)		
Freon 12	Benzene	o-Xylene
Methyl chloride	Carbon tetrachloride	4-Ethyl toluene
Freon 114	1,2-Dichloropropane	1,3,5-Trimethylbenzene
Vinyl chloride	Trichloroethylene	1,2,4-Trimethylbenzene
Methyl bromide	cis-1,3-Dichloropropene	m-Dichlorobenzene
Ethyl chloride	trans-1,3-Dichloropropene	Benzyl chloride
Freon 11	1,1,2-Trichloroethane	p-Dichlorobenzene
Vinylidene chloride	Toluene	o-Dichlorobenzene
Dichloromethane	1,2-Dibromoethane	1,2,4-Trichlorobenzene
Trichlorotrifluoro ethane	Tetrachloroethylene	Hexachlorobutadiene
1,2-Dichloroethane	Chlorobenzene	
cis-1,2-Dichloro ethylene	Ethylbenzene	
Chloroform	m,p-Xylene	
1,2-Dichloroethane	Styrene	
Methyl chloroform	1,1,2,2-Tetrachloroethane	

BULK SAMPLING

Bulk samples may be taken for comparison with the air samples. Sometimes they are gases in liquid. Sometimes they are liquid. Other times they are in powder form.

This process is fairly simple. Identify the sample material, and determine if the sample is suspended in water. If in its concentrated, pure form and not suspended in water, a small sample may be taken by using a precleaned, glass container of about any size in excess of 5 milliliters. Be sure to fill the bottle to overflowing. Carefully install the cap, or slide the separate Teflon into position, cap, and seal the top. This process minimizes air pockets where organic gases may collect. Two samples of each bulk liquid should be taken—one for analysis, the other as a backup.

If diluted or suspended in water, the sample size may have to be coordinated with the analytical laboratory. Generally, a 1 liter sample will be requested where the sample is suspended in water. Fill and seal by the same technique used for smaller samples.

Analysis of the concentrated sample which does not have water can be performed by GC or GC/MS. As part of the air sample "screening process," a bulk liquid sample may be compared against the air samples for retention time on the GC prior to proceeding to the expense of a GC/MS. This may provide an inexpensive means to determine if the associated air sample contaminates originated from the unknown liquid. Following the screening process, the contents of the liquid may be identified by GC/MS instead of, or in tandem with, the air sample. In this fashion, identification and source information may thus be obtained simultaneously.

As for those samples which have been diluted or suspended in water, these materials will likely require extraction prior to analysis. The laboratory will want the diluent identified if possible. Be prepared to provide additional information. After extraction, the samples are processed in a fashion similar to the concentrate.

INTERPRETATION OF RESULTS

The GC/MS has been used to identify chemicals. Once the components are known, they may be analyzed by GC and quantitated more accurately. Those which have been analyzed with the appropriate standard (which is possible with some of the EPA "TO" series) are quantifiable, but without the proper analytical standard, quantitation is not accurate.

Another obstacle is that there are a speculated 20,000 potentially toxic/carcinogenic compounds used in and resulting from chemical processing. Of these, there are only about 400 toxic compounds which have been subjected to extensive epidemiological and animal research. They have assigned acceptable limits, and analytical are readily available to laboratories. Most of the limits (e.g., OSHA permissible exposure limits and ACGIH TWA-TLVs) are for up to 8-hour workplace exposures to the normal adult population. These limits do not apply to those who are elderly or to newly born infants, those who are sickly or susceptible to illnesses. The limits do not apply to 24-hour a day exposures.

Although the EPA had developed acceptable limits for the general population, they are limited in scope even more than the acceptable worker limits. In brief, there are published limits for known chemicals and suggested limits for total organic compounds. The environmental professional attempts to identify all air components while there are abbreviated choices for situational and exposure guidelines and a daunting scope of possible toxins. In response, attempts to do so by others are worthy of comment.

Once properly quantitated, there have been numerous efforts to develop a benchmark for results interpretation of target toxins and their associated compounds. Many of these are based on establishing an acceptable limit are summarized as follows:

- Decide on an acceptable limit for total volatile organic compounds.— Anywhere from 0.05 for any single component to a limit less than 1 mg/m^3 for total components. The total organic limit has been set by some state entities at 0.5 mg/m^3, and some researches suggest levels as low as 0.25 mg/m^3 (as discussed in the section on "Screening Procedures").

- Use the ASHRAE limits—The limits for known toxic substances are one tenth of the ACGIH TWA-TLV.
- Use sensory irritation limits for specific or grouped compounds.

Where limits fail and there has been little or no research performed, comparison sampling is performed. A few of the more frequently used approaches are as follows:

- Compare problem and nonproblem areas.
- Compare indoor and outside air quality.

There is not one singular method which can identify all ambient air components. Formaldehyde and 4-phenylcyclohexane (a suspect irritant in new carpeting) have, on occasion, been identified by GC/MS, but it is unclear as to whether they will be or are always identified or, if identified, the amount present is truly representative of the levels in the ambient air. Further research is needed in order to answer these questions and refine the methods.

The information provided herein has been a synopsis of prevailing thoughts and methodologies as well as a provider of insights as to possible approaches. Although experience in field procedures and laboratory support directs one to the path of least resistance, do not lose site of varying situations, differing needs. Each situation must be considered on its own merits.

Familiarization with known sources and constituents is still the most reliable approach to managing many organic exposure scenarios. Indoor known sources, constituents, and sampling methodologies are discussed in the final chapter.

REFERENCES

1 Kennedy, Eugene, Ph.D. and Yvonne T.G. *Evaluation of Sampling and Analysis Methodology for the Determination of Selected Volatile Organic Compounds in Indoor Air.* (Research document) NIOSH, Cincinnati, Ohio. December 1993.

2 Grote, Ardith A. *Screening Applications Using Thermal Desorption Techniques.* Presented at the AIHC Exposition in Kansas City, Missouri on 23 May 1995. NIOSH, Cincinnati, Ohio.

3 Kennedy, Eugene, Ph.D. and Yvonne T. G. *Evaluation of Sampling and Analysis Methodology for the Determination of Selected Volatile Organic Compounds in Indoor Air.* (Research document) NIOSH, Cincinnatti, Ohio. December 1993. p. 29.

4 Molhave, L. R. Bach, and O.F. Peterson. *Human Reactions to Low Concentrations of Volatile Organic Compounds. Environmental International.* 12:167-75 (1986).

5 Hodgson, Michael, MD, MPH, Hal Levin, B. Arch, and Peder Wolkoff, Ph.D. *Volatile Organic Compounds and Indoor Air. Journal of Allergy and Clinical Immunology.* 2(2):296-303 (1994).

6 Molhave, L. *Volatile Organic Compounds. Indoor Air Quality and Health.* Presented at the Fifth International Conference on Indoor Air Quality and Climate, Toronto, Canada, July 29-August 3, 1990.

7 Meehan, Patrick. *Detection Limits of Ambient Air Samplers.* (Telephone conversation) Environmental Chemistry Division, Texas Research Institute, Austin, Texas, July 1995.

8 Ness, Shirley A. *Air Monitoring for Toxic Exposures.* Van Nostrand Reinhold, New York, 1991. p. 248.

9 Ness, Shirley A. *Air Monitoring for Toxic Exposures.* Van Nostrand Reinhold, New York, 1991. p. 59.

10 Kennedy, Eugene, Ph.D. and Yvonne T. G. *Evaluation of Sampling and Analysis Methodology for the Determination of Selected Volatile Organic Compounds in Indoor Air.* (Research document) NIOSH, Cincinnati, Ohio. December 1993. p. 32.

Chapter 9

CARCINOGENS/MUTAGENS

There are over one million new cases of cancer diagnosed every year in the United States (not inclusive of skin cancers). Approximately half will die of the malignancy.[1] where the same number are thought to be caused by carcinogenic substances in the air and water.[2] Many of the carcinogens have been identified. Yet, the jury is still out on clarifying all cancer-causing agents and their synergy with other substances.

The consensus is that the highest risk pollutants are products of incomplete combustion (e.g., polycyclic aromatic hydrocarbons and their nitrated derivatives). Over half of the air cancer risks in the United States are attributed to vehicle emissions and residential heating while only 20 to 25 percent have been ascribed to industrial emissions. Cancer risk assessments and carcinogen quantitation are, however, the topic of considerable debate.

Cancer risk assessments rely predominately on extrapolations from animal cancer potency data to humans. Some studies overestimate potency while others fail to identify entirely. Although polychlorinated biphenyls are known to cause cancer in laboratory rats, there have been no substantiating epidemiological studies to confirm the same will occur in humans. On the other hand, the drug thalidomide does not cause teratogenesis in rats but has been demonstrated to cause birth defects of children whose mothers had taken the drug. Thus, animal studies alone do not provide substantial proof of chemical impacts on humans.

The complexity of air emissions makes quantitation and identification of "all air- and water-borne carcinogens" even more difficult. The air pollutants rarely remain unchanged. They react or are altered by other chemicals/conditions in their environment. That which was measurable upon discharge loses its identity when mixed with other substances, or the toxicological nature of the mix becomes so complicated that the products are difficult to identify. These potential changes and/or alterations are rarely accounted for by the environmental professional. Yet, they do exist.

Synergism and antagonism between two or more compounds also add to the brew of unknowns. For example, a commonly-accepted synergistic interaction is that of cigarette smoke and asbestos. Separately they are known to result in elevated incidence of lung cancer, yet their combined impact in humans is multiplicative, not additive. Many known/suspect carcinogens have yet to be tested with other substances for synergistic carcinogenicity.

To address these issues, the environmental profession has launched an all-out effort to measure and interpret the sum of "all carcinogenic pollutants" in a given environment. With reserve, the direction has been toward that of "Mutagenicity Testing." The method has been used for environmental pollutant with caution, predominantly by research facilities due to its cost and limited acceptance. Those who dare to go public with their studies have been challenged by their peers and subjected to considerable negative press. Despite these limitations, this method of screening for carcinogens, based upon mutagenicity, deserves mention.

OVERVIEW OF MUTAGENICITY TESTS

A mutagenicity test is a means whereby the mutagenic impact of a substance or mixture of components on mutation-sensitive strains of bacteria can be measured, rapidly and relatively inexpensively (as compared to laboratory animal studies). It is also referred to as the "Ames Test" and the "Salmonella/Mammalian Microsome Liver Test for Detecting Chemical Mutagens."

Table 9.1 Categories of Chemicals Listed as Carcinogens by IARC

Naturally occurring chemicals (e.g., aflatoxins)
Industrial chemicals (e.g., benzidine)
Industrial processes (e.g., chromate-producing industries)
Industrial by-products
Pharmaceuticals

Excerpted from *National Toxicology Program: First Annual Report on Carcinogens.*[3]

Mutagenicity and carcinogenicity are not decidedly one-and-the-same. A mutagen is a substance capable of adding or removing nucleotide bases to and/or from a chromosome, altering the genetic code of a given cell or of a single-celled organism. The result is an altered genetic map which may or may not result in an out-of-control, cellular takeover of a cell or cells which had been designed and programmed to perform specialized functions. Some mutated cells require different nutrients which may or may not be present for cellular maintenance and growth. In the absence of these nutrients, the cell dies. In the presence of these nutrients, the cell may coexist under a different identity.

Those which grow out-of-control, however, could culminate in cancer. When this occurs, the mutagen is considered a carcinogen.

There are a limited number of substances known to cause cancer in man, and many chemicals are merely suspect carcinogens, due to the carcinogenic impact on laboratory animals. Then, there are those which are co-carcinogens which require the presence of other compounds to result in a carcinogen. Physical trauma (e.g., cell damage due to the presence of asbestos fibers), physical agents (e.g., ultraviolet light), and emotional stresses (e.g., psychological distress) are also thought to trigger or initiate cancerous growths. Metabolic changes (e.g., hormonal changes with advancing age) may occur in an organism which will further complicate the chemical stresses. These stresses are neither mutagens nor are they measurable in a majority of the cases.

Table 9.2 Categories of Carcinogenicity

	ACGIH	IARC
Confirmed Human Carcinogens	A1	Group 1
Suspect Human Carcinogens	A2	Group 2
Not Classifiable		Group 3
Probably Not Carcinogenic		Group 4

A substance may be metabolically altered by body chemistry and result in another product which is a mutagen. Without these chemical reactions (variable from one species to the next), the substance may not be mutagenic. Once the substance has been altered in the body, the newly formed mutagen may not have been identified. In this manner, the mutagenic potential of a substance may go unrecognized. These substances may produce negative mutagen test results and go unacknowledged as potential mutagens (or carcinogens) if they have not been metabolically processed. In animal studies, the substance may or may not be metabolized as it is in humans.

In brief, cancer is not attributed, in all cases, to a singular, clearly identified causative agent. A substance which is not mutagenic may become so in certain, as yet unidentified, environments or in tandem with one of the stressors mentioned in the preceding paragraph. Then, too, a single compound may not be mutagenic by itself, but, in combination with another substance or as metabolic by-product, a substance may become carcinogenic. To research all combinations would be tantamount to exploring the galaxies. Accessibility to this knowledge is merely a notion.

ENVIRONMENTAL USES FOR MUTAGENICITY TESTS

When the Ames Mutagenicity Test was first presented in 1975, it was intended as a rapid, inexpensive screening process for carcinogenic substances. The substances which were originally targeted were food additives and chemicals which could have an environmental impact through the food chain. Its use spread to soil and water contaminants and is evolving into a screening or trend monitoring process for its use in tracking mutagens generated by vehicular emissions and industrial activities. These tests are, however, always substantiated and backed by other forms of quantitative monitoring of known carcinogens.

Mutagenicity testing has been used to evaluate water quality. Comparisons have been performed of drinking water, river water, lakes, waste water, creeks, well water, and sewers testing upstream and downstream of industrial effluent and the effluent discharge may be performed. Trends can be developed, and cases involving altered mutagenicity of an effluent after environmental mixing have sparked further studies and identification of the cause.

The observed high incidence of cancer in a population may trigger a study to locate a common source factor. Suspect drinking water (e.g., water wells) may be screened for mutagenicity against other sources as well as commercial bottled water. This may be done, part-and-parcel, in conjunction with attempts to identify known suspect carcinogens through chemical analyses.

Mutagenicity testing provides a quick overview of complex mixtures and their composite mutagenic potential without having to go to the lengthy guessing game of "What is it?" and "Have the individually identified, known mutagens been enhanced by other components of the mix?". The singular impact of known carcinogens may contribute to merely a fraction of the mutagenic potential. The mutagen test allows for all known and unknown components to be measured.

The mutagenic potential of rural air has been compared with that of industrial air, and seasonal/diurnal variations have been simultaneously recorded. Such time-related studies have been used to track down periods when mutagenic air components were elevated, allowing for a correlation between mutagenic contributions and activities. For example, the likelihood for elevated mutagenic air components in New Jersey is greatest when the refineries are most active and during peak vehicular traffic periods. These have been effectively tracked.

In one study, the mutagenicity of the air in a town located near oil field activities was compared with that of a nonindustrial city and a rural town. The results indicated "many more revertant colonies" in the oil field town over that of the nonindustrial city. The nonindustrial city had more than the unexposed blank, and the rural location had no demonstrable revertants.

Attempts have been made to monitor fire by-products as well. In a chemical fire, relative mutagenicity of the components downwind, particle/gas laden air verses the upwind, uncontaminated air may provide information sufficient to confirm or deny the need for increased levels of carcinogen exposure concerns. In one study, the mutagenicity of the smoke from a burning petroleum tower was greater than the sum total of the mutagens which were identified individually.

Mutagenic potential has also been found to vary depending upon the burn temperature where waste is incinerated. The changes which result are complex, but they can be easily classified as to their mutagenicity without the need for extensive chemical analyses.

The issue of carcinogenicity and mutagenicity has arisen within the list of concerns in indoor air quality studies. Tobacco smoke, formaldehyde, volatile solvents, asbestos, and radon are only a portion of the list of potential carcinogens, and these materials are trapped, retained more readily than the outdoor air environmental substances. Oftentimes, building occupants express the concern for mutagenic substances, and the environmental professional typically targets specific components. The wherewithal for mutagen sampling has been discussed. Yet, there have been minimal publications as to findings.

Laboratories have attempted to promote the use of the Mutagenicity Test, but awareness has been limited. Mutagenicity assay costs are two to three times that of waste stream identification analytical costs, and there are no standard methods for sampling and interpretation of the results. Nevertheless, comparative sampling may be performed much the same way as is done for many of the other unknowns, and with increased usage, advances in automation, the costs may become more feasible.

In brief, the mutagenic potential of a complex mix of water, dust, and air may be evaluated to screen for risk. Once screened, the implicated material or materials may be subjected to additional chemical analyses (e.g., polynuclear aromatic compounds). Until Mutagenic Testing is commonly accepted by environmental professionals, confirmation analyses are recommended for both negative and positive mutagenic results.

SAMPLING METHODOLOGIES

The most common substances sampled for and analyzed by the Mutagenicity Tests have been pesticides, industrial chemicals, and surgical supplies. These singular, uncomplicated materials have been the primary targets when performing rapid "screening for mutagenicity." In environmental and occupational studies, however, sampling is performed primarily of complex mixtures of water, surface dust, and air which typically are analyzed for chemical composition (which must be known or suspect). Where the

Mutagenicity Testing is performed, however, the composition does not need to be known nor suspect in order to perform the analysis, and the sample collection methods are relatively simple.

Water sampling involves the collection of 1 to 2 gallons of water in a pre-cleaned, amber glass bottle or bottles, and the lid should be made of a nonreactive substance (e.g., Teflon-lined cap). Although the analytical method generally requires extraction from a 2 liter sample, larger quantities of collected water provide an excess of sample material should additional analyses be indicated. Overfill the container into which the water is being collected. Upon placement of the lid, slide the cover over the raised overflow sample to avoid retaining an air pocket within the bottle. Secure the lid, and seal around the edges. Label the contents, place in cold storage, and ship to the laboratory.

Surface dust sampling involves the collection of a minimum 2 milligrams of "representative dust." Care should be taken to assure there is no cross contamination between samples or from other sources (e.g., hands), and the sample material should be placed in a clean, nonreactive container (e.g., a glass container with a Teflon-lined lid). Label the contents, and ship to the laboratory.

Air sampling methods are not well defined, but the requirement remains for a 2 milligram sample. In order to collect a sufficient sample of representative dust from the air, extremely large volumes of air must pass through a pre-cleaned, nonreactive filter material (e.g., glass fiber filter which has been heated to 320° C for 24 hours or longer and stored, wrapped in pre-cleaned aluminum foil). The suspended material is then collected on the surface of the filter.

Gravimetric dust sampling of outdoor environments often result in environmental dust levels less than 0.05 mg/m^3. Based on 0.05 mg/m^3, the environmental professional must sample a minimum of 40 cubic meters (or 40,000 liters) of air in order to collect a 2 milligram dust sample. This is air with no visible pollutants.

A study involving particulate sampling in a firefighting smoke plum indicated dust levels as high as 300 mg/m^3 which would require less than 700 liters of sampled air volume. Indoor dust levels vary. Yet, indoor dust levels may be as low as 0.025 mg/m^3 requiring a 80 cubic meter air sample. The minimum target air sample volume must be adjusted according to conditions and activities monitored. In one study, the target sample volume was standardized at 75 cubic meters of air which allows for minimum and provides additional material which may be used for other analyses. Whichever target air volume is chosen, the environmental professional will need to be able to sample at a rate in excess of 100 liters per minute.

Sample duration enters into the overall picture in determining flow rate. If a particular time period is limited by activity to a 1 hour sample, the required flow rate may be 350 liters per minute (for visibly clean outdoor air) or 700 liters per minute (for indoor air). Traditional industrial hygiene high volume sampling pumps do not have the capacity to meet these demands. Thus, environmental particle sampling equipment (which range from 100 to 1,000 liters per minute) is recommended. Once the sample duration has been decided, determine the minimum flow rate to collect the desired air volume. Set the flow rate on the vacuum pump. Install an appropriately sized, nonreactive filter. Record the sampling start time and finish time. Calculate the total air volume samples, and forward the air volume information and filter to the analytical laboratory in a clean, nonreactive enclosure (e.g., aluminum foil packet).

ANALYTICAL METHODOLOGY

Mutagenicity testing is based upon the supposition that if a substance causes mutations in bacteria it will also cause mutations in humans. In the 1970s, Dr. Bruce Ames validated his test by showing that 90 percent of 175 organic chemical carcinogens tested positive for mutagenicity. There have since been various attempts to correlate mutagenic findings of specific chemicals with that of animal exposure studies and carcinogenicity. See Table 9.3 for target carcinogenic substances and mutagenicity test results on those chemicals tested and their results published. Not all chemicals tested have been published, and there is no common repository for this information.

In 1993, a review of mutagenicity and carcinogenicity data of 384 chemicals indicated that 34 percent of those substances which tested positive for carcinogenicity (i.e., rat and mouse tests) also tested positive for mutagenicity (i.e., *Salmonella* test). Only 10 percent of the total tested negative for carcinogenicty and positive for mutagenicity, and mutagenicity tests missed 45 percent of the total numbers which tested positive in carcinogenicity tests. In brief, the test results agreed in 72 percent of the cases (positive and negative concurrence). These are summarized in Table 9.4.

Various strains of *Salmonella typhimurium* bacteria are mutated so they can no longer make the amino acid histidine which is required for growth. About one billion histidine-requiring bacteria are plated onto a Petri dish containing nutrient agar with limited amounts of histidine to allow for minimal growth in the presence of the test substance and homogenized rat liver tissue extract. The inclusion of a rat liver extract simulates metabolic processing of the test substances. Human liver extract would be ideal in mutagenicity tests, much the same as human test subjects would be for toxicity tests. However, this approach is impractical for routine testing due to the limited availability of human liver. Also, chemical processing does not always occur in the liver, and not all individuals have the same chemical make-up nor do individuals metabolize substances the same. So, the method has attempted simulation within the bounds of feasibility.

Table 9.3 American Conference of Governmental Industrial Hygienists (ACGIH) List of Carcinogens vs. Ames Mutagenicity Test Results

Chemical	Ames Test Results
KNOWN TO CAUSE CANCER IN HUMANS (1995)	
4-Aminodiphenyl	+
Arsenic, elemental and inorganic (except arsine)	
Asbestos	
Benzene	-
Benzidine	+
Chromite ore processing (chromate)	
Chromium (VI) compounds	+
Coal tar pitch volatiles [benzene solubles; same as particulate polycyclic aromatic hydrocarbons]	+
beta-Naphthylamine	+
Nickel, elemental	
4-Nitrodiphenyl	
Vinyl chloride	+
Zinc chromates	

* *Italicized substances are considered definite carcinogenic substances by the International Agency for Research on Cancer as well as the ACGIH.*

Chemical	Ames Test Results
SUSPECT HUMAN CARCINOGENS	
Acrylamide	-
Acrylonitrile	+
Antimony trioxide [production]	
Benz(a)anthracene	+
Benzo(b)fluoranthene	
Benzo(a)pyrene	+
Beryllium and compounds	
1,3-Butadiene	+
Cadmium (elemental and compounds)	
Calcium chromate	
Chloroform	+
bis(Chloromethyl)ether	
Chrysene	
3,3'-Dichlorobenzidine	+
Dimethyl carbamoyl chloride	
1,1-Dimethylhydrazine	
Dimethyl sulfate	
Dinitrotoluene	+
Ethyl acrylate	-
Ethylene oxide	+
Formaldehyde	+

Table 9.3 (continued)

Hexachloroethane	-
Hexamethyl phosphoramide	
Hydrazine	
Lead chromate	
Methyl hydrazine	
4,4'-Methylene bis (2-chloroaniline) [MOCA]	+
4,4'-Methylene dianiline	+
Methyl iodide	
2-Nitropropane	
N-Nitrosodimethylamine	+
N-Phenyl-beta-naphthylamine	
o-Phenylenediamine	
Phenylhydrazine	+
Propane sultone	+
beta-Propiolactone	+
Proopylene imine	
Strontium chromate	
o-Tolidine	
o-Toluidine	+
p-Toluidine	+
Xylidine	

Excerpted from *ACGIH List of Carcinogens* and "Comparison of target organs of carcinogenicity for mutagenic and nonmutagenic chemicals." [4,5]

Typically, the test substance must be extracted from water/dust samples prior to application onto the dish. The inoculated Petri dish is then incubated at body temperature (i.e., 37° C) for two days. In the presence of a mutagen, some or many of the histidine-requiring bacteria will revert back to histidine producer and be able to grow in the restricted nutrient agar. These growths are visibly identified as colonies. Thus, each colony represents a single bacterial revertant. The greater the number of colonies, the greater the number of revertants, or mutations. Most mutagens are detected at very low levels, in the nanogram amounts. For governmental requirements, four different strains are tested. Environmental studies, however, have no ready source of prerequisites, so the two most sensitive stains are generally used.[6] Since inception, there have been numerous strains developed and refinements to the process. Although there is an effort underway to automate the method, the newer Mutagenicity Tests may not result in cost reductions.

Table 9.4 Comparisons of Mutagenicity in Salmonella and Carcinogenicity in Rats and Mice (384 Chemicals)

		Carcinogenic +	Carcinogenic −
Mutagenic	+	131 (34% of total)	38 (10% of total)
	−	106 (28% of total)	109 (28% of total)

INTERPRETATION OF RESULTS

In the environmental profession, there are neither regulatory nor recommended acceptable/nonacceptable limits for Mutagenic Testing. The test should not be used as means to confirm or deny carcinogenicity. It may, however, be used as a pre-screening tool to identify specific areas which may require more in-depth sampling, or it may be used to track trends of commonly accepted agreed-upon, previously identified carcinogens. Proceed with caution!

REFERENCES

1. American Cancer Society. "Cancer Facts and Figures–1995." [Bulletin] American Cancer Society, Atlanta, Georgia, 1995.
2. Flessel, Peter, Yi Y. Wang, Kuo-In Chang, and Jerome J. Wesolowski. Ames Testing for Mutagens and Carcinogens in Air. *Journal of Chemical Education*, May 64(5):391-5 (1987).
3. National Toxicology Program. First Annual Report on Carcinogens. [Bulletin] Department of Health and Human Services, Volume 1. (July 1980).
4. American Conference of Governmental Industrial Hygienists. *List of Carcinogens.* ACGIH, Cincinnati, Ohio. (1995)
5. Gold, Lois Swirsky, et. al. Comparison of target organs of carcinogenicity for mutagenic and non-mutagenic chemicals. *Mutation Research.* 286:75-100 (1993).
6. Ames, Bruce N. *Environmental Chemicals Causing Cancer and Genetic Birth Defects: Developing a Strategy to Minimize Human Exposure.* Institute of Governmental Studies, University of California, Berkeley, California, 1978.

Chapter 10

PRODUCT EMISSIONS

Indoor nonindustrial exposures to organic compounds are typically two to one hundred times higher than those found outdoors. This is primarily attributed to emissions from construction materials, furnishings, office supplies/equipment, maintenance and cleaning products. Other contributing sources include individual use products (e.g., perfumes and lighters) and outdoor air pollutants.

Indoor industrial exposures to volatile organic compounds are generally ten to one hundred times those found in nonindustrial environments. This extreme difference raises a jaundiced-eye whenever an environmental professional responds to office building complaints. Exposures are so much lower in nonindustrial environments. Yet, complaints are more prominent.

The problems/complaints in office buildings appear disproportionate to the lack of problems/complaints in industry. Some feel that the discrepancy lies in the medley of chemicals. Office environments generally consist of up to three hundred chemicals, an amalgam that far exceeds most industrial exposures that might consist of up to ten components. So, due to complexity of building construction/furnishing emissions, architects, construction managers, building owners, state agencies, and other organizations have begun to seek supplies with "minimal product emissions."

Space management plans have taken into consideration the building design, number of occupants, and location of occupants. Yet, as less space becomes available, crowding culminates in more congestion, greater product emissions.

Renovation activities, pesticide treatments, maintenance and cleaning products contribute, and outdoor repairs/roofing activities contribute to the quality of the indoor air. Some construction materials and many furnishings concentrate or amplify airborne chemical contaminants that are introduced by other materials or treatments due to cleaning products, pesticide applications, and maintenance activities. Whereas they may not be a direct source of exposure, these materials may contribute indirectly.

Construction practices and building commissioning/recommissioning are also included in the plethora of issues that impact the burden of airborne chemicals. New building construction and recently renovated indoor environments have been found to have volatile organic chemical levels similar to those found in industry (e.g., up to 30 mg/m^3 of volatile organic compounds).

Surprisingly, the higher levels are associated with residential structures more than those found in office buildings.

On an increasing frequency, products are being selectively manufactured to minimize or eliminate emissions all together in nonindustrial indoor environments. Market awareness is the driving force. Product emissions testing is the first line of defense!

USES FOR PRODUCT EMISSIONS TESTING

Product emission testing is used by manufacturers for product information and by environmental professionals for "predicting product emission contributions" to indoor air environments. Although the ultimate concern is product emissions (frequently referred to as off-gassing) in new or renovated office environments after installation, a more practical approach is required testing prior to any significant purchases.

Table 10.1 Components Which Result in Product Emissions[1]

VOLATILE ORGANIC COMPOUND EMISSIONS		
	Paints	
	Fabrics and fabric treatments	
	Cushions (polyurethane/polystyrene foam and polyester stuffing)	
	Plastics (various plasticizers)	
	Adhesives	
	Cleaning solvents	
FORMALDEHYDE EMISSIONS		
	Particleboard	Glues
	Plywood	Resins
	Pressboard	Insulation
	Paneling	Foam
	Carpeting/carpet backings	Laminates
	Dyes	Plastics/moldings
	Household cleaners	Stiffeners
	Wrinkle-resistant fabrics	Water repellents

Testing is performed by some manufacturers for quality control. The manufacturer may have testing performed at various stages in the production process to determine the 24-hour emission factor for a predetermined time period. This is generally performed, if not intermittently, upon the finished product. With predetermined limits, the manufacturer addresses and corrects manufacturing discrepancies to minimize the sale of products that may contribute excessive irritant/hazardous substance levels to the indoor air environment. The manufacturer, also, uses this information to identify process variances which may contribute to elevated emissions from their products. In this manner, manufacturing processes are altered to accommodate the need for reduced off-gassing. Products which frequently receive considerable attention are particleboard (e.g., formaldehyde) and carpeting (e.g., 4-phenylcyclohexene). Some other products implicated in product emissions are listed in Table 10.1.

Product testing is performed to meet special labeling restrictions. The Carpet and Rug Industry has a pass/fail indoor air quality labeling program for its participants. Upon successfully passing the emissions test, the participant is approved to place a green and white label on their carpeting that states "Carpet and Rug Industry Indoor Air Quality Testing Program."

Carpet Component: ***4-phenylcyclohexene (binder, backing, and glue)***

Copiers and laser printers also have a labeling program. It is referred to as the "Blue Angel." Only products meeting strict emissions limits may be labeled. Although the labeling originated in Germany, there is a growing interest internationally.

Some environmental professionals address product emissions prior to construction and purchase of furnishings. The State of Washington has instituted an East Campus Plus Indoor Air Quality Program of all state-owned buildings. It involves construction practices, selection of materials, methods for commissioning renovated/newly constructed buildings, and acceptable office building limits for total volatile organic compounds and formaldehyde. The Program requires all major construction materials and office furnishings to undergo environmental chamber testing and, based on predicted levels, ensure the products will not, in combination, exceed the acceptable limits. Any products exceeding the limits are required pre-installation airing. The means used for predicting the total burden is through an EPA computerized algorithm with emission rates and other building specific features (e.g., volume, air changes, etc.).

Toxicological testing which is based on environmental chamber technology and animal inhalation studies is performed to address sensory irritation caused by airborne chemicals. Product emissions are introduced into a chamber housing laboratory animals. The breathing patterns and other clinical responses are then monitored. Challenges are singular or multiple.

There have been isolated instances of human irritation testing using environmental chambers. For example, a researcher who has received considerable attention internationally used chamber challenge testing to develop a dose response relationship for discomfort due to volatile organic compounds. See Table 10.2 for a summary of his findings.

Table 10.2 Summary of Research Findings on Effects of Total Volatile Organic Compound (TVOC) Mixtures[2]

TVOC (mg/m^3)	Health Effects/ Irritancy Response
<0.20	No response
0.20 - 3.0	Irritation and discomfort
3.0 - 25	Discomfort (probable headache)
>25	Neurotoxic/health effects

On this order, there have been attempts by some researchers to perform challenge testing without the benefit of a chamber and/or known, monitored levels of exposure. As the actual chemical exposure levels are poorly controlled, the results become questionable. Furthermore, there is a concern for life threatening patient reactions, and a medical emergency must be considered when performing the tests. For these reasons, human challenge testing is rare in the United States.

MEASURING UNITS AND EXAMPLES

Product emissions and their impact on a confined office/building environment are measured by one of several approaches, each with an emphasis on and targeting different information. A few of the more commonly used units include emission factors, emission rates, and predicted air concentrations.

An "emission factor" is the amount of chemical which is emitted from a product at a specified time in the life of the product.[3] For example, a product may be tested immediately after production, just prior to shipment, or prior to installation. The time is "defined," and numbers vary during the manufacturing evolution and life of a product. Solid product emissions are based on exposed surface area(s), micrograms per square meters-hour (μg/m^2-hour). Liquid products may be based on relative mass units, micrograms per gram-hour (μg/g-hour), and whole units may be reported in composite form, micrograms per composite-hour (μg/composite unit-hour). In the latter, the composite may be a single workstation or a group of furnishings. An example follows:[1]

System office furniture

- Formaldehyde: 802 to 3780 mg/workstation-hour
- Volatile organic compounds: 160 to 45,000 mg/workstation-hour

Chairs

- Formaldehyde: no detection to 1670 mg/chair-hour
- Volatile organic compounds: 159 to 450 mg/chair-hour

Tackable acoustical partitions (with phenol-formaldehyde treated fiberglass insulation)

- Formaldehyde: 0.158 to 0.37 mg/m^2-hour
- Volatile organic compounds: 0.006 to 0.074mg/m^2-hour

See Table 10.3 for additional findings. Each of these studies is limited in scope therefore not to be considered conclusive and/or all encompassing regarding all products. Variations will occur between product types, manufacturers, formulations, and production processes.

Table 10.3 Emissions from Furnishings/Related Products[1&4]

Product	Emission Rate (µg/m2-hour)
TOTAL VOLATILE ORGANIC COMPOUNDS	
Solvent-based adhesives	up to 17,000,000
Water-based adhesives	up to 2,100,000
Furniture spray polish	300,000
Wood stain	17,000
Polyurethane lacquer	6,000
Plywood	up to 2,400
Polystyrene foam	up to 1,400
Particleboard-fiberboard	up to 150
Hardboard	30
Medium density fiberboard	40
FORMALDEHYDE	
Wood products	170 to 900
Insulation	16 to 26
Wall coverings	20 to 600
Textiles	not detected to 3,000

Table 10.4 Formaldehyde Emission Factors from Finished Wood Products[1]

Material	Concentrations (µg/m^3)
Medium density fiberboard	970
Unfinished particleboard	up to 809
Finished particleboard	up to 719
Finished medium density fiberboard	up to 246
Water-damaged chipboard	48
Hardboard	up to 30
Medium density fiberboard	up to 14
Non-water-damaged chipboard	10

"Predicted air concentrations" are location specific. Calculations are based on the product's "emission rate" as well as specific building environmental components.[3] These components include air movement, amount of make-up air, air volume capacity of the room(s), and level of occupant activity. The complexity of the algorithm lends itself to computer modeling. The EPA Exposure Model is widely used, and its predictions have been validated.

SAMPLING METHODOLOGIES

There are no hard, fast rules for sampling. Indoor sources of emissions vary widely in both the strength of their emissions and in the type/number of compounds emitted. Differences in the emission rates vary to several orders of magnitude even with the same type of material. So, the amount of material required for analysis will vary by the experience of the commercial laboratory and their directives on a case-by-case situation.

Then, too, entire modules may require special handling practices, and a large test chamber that limits the choice of laboratories that are capable of performing the analytical process. The environmental professional should be wary of the time the sample is extracted or packaged, precautions for containment, and transport environment when planning to ship a sample for analysis. All this should be arranged and coordinated through the laboratory of choice prior to sample collection.

ANALYTICAL METHODOLOGY[5]

Although there is no standard protocol for environmental chamber emissions testing, the American Society for Testing and Materials (ASTM) published a *Standard Guide for Small-Scale Environmental Chamber Determinations of Organic Emissions from Indoor Materials/Products* in November 1990. They clearly state that this is a guide only and differences in approach will occur from one researcher/laboratory to the next. A distinction is made between small and large chambers. Small chambers (smaller than 5 cubic meters in volume) are not to be used to evaluate applications (e.g., spray painting activities).

Products (e.g., small samples or whole furnishing modules) are placed in or material (e.g., paint) applied inside a contained, controlled test chamber. The humidity, temperature, chamber air exchange rates, and air movement are controlled/constant. Air monitoring is performed at various predetermined intervals at the "headspace" (e.g., air immediately above the sample) of either closed containers or dynamic flow-through headspace. Collection is performed for a minimum of the following:

- Volatile organics
- Semivolatile organics
- Polar compounds
- Nonpolar compounds

Monitoring methods may involve capture using a thermal desorption tube or other media as required for that material which is of interest. Analyses are typically performed by GC/MS.

INTERPRETATION OF RESULTS

In the absence of Federal standards for product emission limits, there are a few industry guidelines, researcher recommended guidelines, and occasional state/private building owner standards.

As indoor air quality concerns increase and the public expresses greater interest in product emissions, more industries are expected to set their own criteria for products with low emission rates. The carpet industry was one of the first to show an interest. In order to qualify for the Green Label, carpets must not exceed specified emission factors as measured over a 24 hour exposure period. These emission factor limits are:[6]

- Volatile organic compounds: 0.5 mg/m^2-hour
- Styrene: 0.4 mg/m^2-hour
- Formaldehyde: 0.1 mg/m^2-hour
- 4-Phenylcyclohexene: 0.05 mg/m^2-hour

In Germany, environmental labeling focuses on the environmental acceptability of indoor products, based on chemical emissions and other environmental concerns (e.g., waste generation and recycling). A "Blue Angel" emissions criteria is provided for copiers and laser printers. On the international market, manufacturers are perceiving the growing significance of emissions testing and seeking Germany's "Blue Angel" label for their copiers and laser printers. The maximum allowable emissions are based on environmental contributions and must not exceed the following:[7]

- Dust: 0.25 mg/m^3
- Ozone: 0.04 mg/m^3 (0.02 ppm)
- Styrene: 0.11 mg/m^3 (0.025 ppm)

A researcher with the Environmental Protection Agency suggested certain product "default values" wherever predicted air modeling cannot or is not going to be performed. These values are for volatile organic compounds only and are based on emission rates:[8]

- Flooring materials (including carpeting): 0.6 mg/m^2-hour
- Wall materials: 0.4 mg/m^2-hour
- Moveable partitions: 0.4 mg/m^2-hour
- Office furniture: 2.5 mg/workstation-hour
- Office machines (central): 0.25 mg/hour-m^3 of space
- Office machines (personal office): 2.5 mg/hour-m^3 of space

The State of Washington has what is referred to as the "East Campus Plus Indoor Air Quality Program." This sets criteria for construction practices, materials selection, and building commissioning as well as maximum allowable concentrations to various substances in indoor air environments. The permissible limits for airborne composite sampling are: [8]

- Volatile organic compounds: 0.5 mg/m^3
- Total particles: 0.05 mg/m^3
- Formaldehyde: 0.06 mg/m^3 (0.05 ppm)
- 4-Phenylcyclohexene (carpet only): 0.0065 mg/m^3

This Program also requires all major construction materials and office furnishings to undergo environmental chamber testing to ensure the products will not, in combination, exceed the permissible limits. Products that are determined, through chamber testing, potential contributors in exceeding the permissible limits require pre-installation airing out prior to placement within a building for which predicted air modeling has been performed.

The State of Alaska has adopted these criteria for new construction projects, and California Proposition 65 requires products be labeled if emissions exceed defined limits. At the present, most of the state limits are principally for new construction on state-owned buildings. Privately owned buildings are internally managed with little or no guidelines.

REFERENCES

1 Franke, Deborah, et. al. Furnishings and the Indoor Environment. [Draft to be published in the *Journal of the Textile Institute*] Air Quality Sciences, Inc., Atlanta, Georgia, 1995.

2 Molhave, L. "Irritancy of Volatile Organic Compounds in Indoor Air Quality." Paper presented at the Fifth International Conference on Indoor Air Quality and Climate in Toronto, Canada, 1990.

3 Air Quality Sciences. "Defining Product Emission Measurements." [Bulletin] Air Quality Sciences, Inc., Atlanta, Georgia, 1995.

4 Black, Marilyn, Ph.D. "Volatile Organic Compounds in the Indoor Environment." [Bulletin] *Indoor Environment '95*, May 1-3, 1995.

5 American Standard and Testing Materials (ASTM). *Standard Guide for Small-Scale Environmental Chamber Determinations of Organic Emissions from Indoor Materials/Products.* ASTM D5116-90. 1990.

6 Air Quality Sciences. How Do CRI's Carpet Emissions Criteria Compared To The State of Washington's Purchase Specifications? [Newsletter] *AirfAQS*, Air Quality Sciences, Inc., Atlanta, Georgia. Fall 2(1):4 (1994).

7 Air Quality Sciences. Office Machine Manufacturers Seek Germany's Blue Angel Certification. [Newsletter] *AirfAQS,* Air Quality Sciences, Inc., Atlanta, Georgia. Winter 2(2):1–2 (1995).

8 Air Quality Sciences. What Guidelines Exist for Chemicals and Particles? [Newsletter] *AirfAQS*, Air Quality Sciences, Inc., Atlanta, Georgia. Summer 1(4):2 (1994).

GLOSSARY

abscess—A cavity containing pus and surrounded by inflamed tissue.

acid-fast—A method of bacterial identification. Acid-fast bacteria retain fuschin stain where other bacterial are rapidly decolorized when treated with a strong mineral acid.

adenine—A DNA base which pairs to thymine (on DNA) or uracil (on RNA).

adsorption—Process of collecting a liquid or gas onto the surface of a solid or liquid sorbent.

aerobic—Requiring the presence of oxygen for growth. Some aerobic microbes may form capsules, or spores, when left in an oxygen deficient environment.

aflatoxin—Mycotoxin created by the molds *Aspergillus flavis* and *parasiticus,* a known cancer-causing substance.

algae—Plant-like organisms that practice photosynthesis (requiring light) and, for the most part, live in aquatic environments. They are occasionally implicate with outdoor allergies.

allergic pneumonia—An inflammation of the lungs resulting directly from an allergy, usually to some type of organic dust.

alveolitis—An allergic pulmonary reaction characterized by acute episodes of difficulty breathing, cough, sweating, fever, weakness, and pain the the joints and muscles, lasting from 12 to 18 hours.

amoeba—A protozoan which has an undefined, changeable form and moves by pseudopodia (or branching fingers of cellular material).

amplification—In reference to microbes, this means an increase in the number.

anaerobic—Not requiring the presence of oxygen for growth (e.g., *Clostridium botulism* which causes canned food poisoning).

angioedema—This is a condition similar to urticaria (or hives), but involves the subcutaneous tissue. It ordinarily does not itch and is a more generalized swelling.

anisotropic—Substances having different refractive indices which depend on the vibration direction of light.

anneal—A process of heating and cooling which in the PCR process involves a cooling process to cause the primers to match with the DNA strands or duplicated primers.

antiparallel—Complementary strands of DNA are antiparallel.

aqueous waste—Waste that is greater than 50 percent water.

ascospore—A haploid sexual spore created by the fungal class Ascomycetes (e.g., yeasts).

asthma—A disease that is characterized by recurrent episodes of difficult breathing and by wheezing, with periods of nearly complete freedom from symptoms.

atopic—Of or pertaining to a hereditary tendency to develop immediate allergic reactions, such as allergic rhinitis, allergic asthma, and some forms of eczema.

auto immune disease(s)—A person's immune system reacts against its own tissues and organs.

bagassosis—A form of allergic lung disorder caused by exposure to moldy sugar cane fiber.

bake-out—A process whereby the temperature of a building is elevated to force chemical off-gassing of building materials and furnishings. This is generally performed without building occupants, and the air is flushed with outside air after a predesignated time period.

basidiospores—Sexual spores created by the fungal class Basidiomycetes (e.g., common mushroom).

Becke line—A bright halo which is observed near the boundary of a particle that moves with respect to that boundary as the microscope is focused.

birefringence—The numerical difference in refractive indices for a substance.

booklice (or book louse)—Small, six-legged insects belonging to the order Psocoptera which are implicated in paper-dust allergies.

bronchitis—An inflammation of the mucous membranes of the bronchial tubes, characterized by difficulty breathing***carcinogenic***—A substance capable of causing cancer.

challenge test—A medical procedure, also known as provocative testing, used to identify substances to which a person is sensitive by deliberately exposing the person to diluted amounts of the substance. A positive bronchial challenge is one in which pulmonary function decreases.

commensal—Organisms that live in close association whereby one may benefit from the association without harming the other.

competition—Negative relationships between two populations in which both are adversely affected with respect to their survival and growth.

conidia—Asexually-produced spores which developed at the end of a conidiophore (e.g., *Penicillium*).

conidiophore—The structure from which conidia develop.

conjunctiva—The mucous membrane lining of the inner surfaces of the eye.

conjunctivitis—Inflammation of the conjunctiva. This results in eye redness, a thich discharge, sticky eyelids upon waking in the morning, and inflammation of the eyelids.

croup—An infection of the upper and lower respiratory tract that occurs primarily in infants and young children up to 3 years of age. It is characterized by hoarseness, fever, a distinctive harsh, brassy cough, a high pitched sound when breathing, and varying degrees of respiratory distress.

culturable—Viable microorganisms which can be grown.

cytosine—A DNA base which only pairs to the base guanine.

deoxynucleotides—Genetic information, includes A (adenine; deoxyadenylate), T (thymine; deoxythymidylate), G (guanine; deoxyguanylate), and C (cytosine; deoxycytidylate).

deoxyribonucleic acid—Singular or paired strands of deoxynucleotides; double strands form a helix; in double strands, one DNA is complementary to the other.

deoxyribose—Sugar unit associated to a nucleotide, or DNA link.

desorption—Process of running adsorbed liquids/gases from sorbent material.

dideoxy sequencing—A genetic sequencing process which uses DNA polymerase, a DNA template, nucleotide triphosphates, and dideoxynucleotide triphosphates (ddNTP's).

dideoxynucleotide triphosphates—Nucleotides which will terminate the addition of other nucleotides to a gene sequenced template; a "cap" to or termination of the sequencing.

dispersion staining—Staining based upon the differences and similarities of the refractive indices between a solid and liquid.

dispersion—The separation of light into its color components by refractive or diffracted light.

DNA—Deoxyribonucleic acid, or the genetic material which dictates cellular activities.

dust mites—Four-legged arachnids which are typically implicated in house dust allergies.

dyspnea—Difficulty breathing.

eczema—A superficial dermatitis which, in the early stages, is associated with itching, redness, fluid accumulation, and weeeping wounds. It later becomes crusted, scaly, thickened, with skin eruptions.

eczema—An inflammation of the skin, marked by redness, itching, and scales.

emission factor—A single point quantitative measurement of gaseous or particle emissions from a material, as determined by chamber testing.

emission rate—The actual rate of release of vapors/gases from a product over time.

emphysema—A chronic disease of the lungs, in which the alveoli are permanently damaged or destroyed. It is typically characterized by difficult breathing and is associated with heavy, prolonged cigarette smoking.

endocarditis—Lesions of the lining of the heart chambers and hear valves.

endotoxic shock—A body reaction caused by an endotoxin, generally characterized by marked loss of blood pressure and depression of the vital processes.

endotoxin—A toxin produced within bacteria and released upon destruction of the cell in which it was released.

environmental chamber—A nonreactive testing enclosure of known volume with controlled air change rates, temperature, and humidity.

epitope—The portion of an antigen molecule involved in binding to the antibody.

extinction—Where the orientation of crystals which appear white or colored between crossed polars is changed by rotating the stage, all single crystals will seem to disappear as they darken; this reveals the vibration directions of crystals.

farmer's lung—A form of allergic lung disorder caused by exposure to moldy hay.

flush out—Remove contaminants from a contained air space.

fomites—Inanimate objects (e.g., mineral dust particles).

forensic—Relating to or dealing with the application of scientific knowledge to legal issues.

fungi—Plants that, unlike green plants, have no chlorophyll and must depend on plant or animal material for nourishment.

gastroenteritis—An inflammation of the stomach and intestines. It may at times result from an allergic reaction.

gene—Section of DNA that codes for one protein or enzyme.

Gram negative—A staining process which is used in diagnostic bacteriology whereby the bacterial wall is stained pink.

Gram positive—A staining process which is used in diagnostic bacteriology whereby the bacterial wall is stained purple, almost exclusively a property of producers of potent exotoxins.

granuloma—A granulated nodules of inflamed tissue.

guanine—A DNA base which only pairs to the base cytosine.

hapten—A low molecular weight chemical that is too small to be antigenic by itself but can stimulate the immune system when combined with a larger molecule (e.g., protein).

hay fever—Common name for "nasal allergy." Its symptoms include attacks of sneezing, runny, stuffy nose, and itchy, watery eyes, and they occur within a few minutes to a few hours after exposure to inhaled allergens—usually pollen, spores or molds, house dust, or animal dander. The term hay fever is misleading, since these reactions are not usually produced by hay and are not accompanied by fever.

hemorrhagic shock—Physical collapse and prostration associated with sudden and rapid loss of large amounts of blood.

heterologous (cross-reactive) antigen—A different antigen from that which was used to immunize or challenge the immune system, yet it is similar enough to be recognized as the initial foreign substance. Typically, these antigens are polysaccharides, possibly due to their limited chemical complexity and often structurally similar nature to one another. For example, human blood Group B sometimes reacts with antibodies to certain strains of *Escherichia coli* which is a common bacterial resident in the human colon.

hives (urticaria)—A common skin condition that is probably familiar to everyone. It is a skin rash characterized by areas of localized swelling, usually very itchy and red, and occurring in various parts of the body. It usually lasts only a few hours and involves only the superficial areas of the skin. Urticaria has either an immunologic or a nonimmunologic cause.

homologous antigen—A foreign substance used in the production of antiserum.

house dust—Heterogeneous, firm gray powdery material that accumulates indoors. This category includes mold, pollen, animal dander, food particles, kapok, cotton lint, insects, and bacteria.

hybridized—Bound to a template of DNA sequencing.

hypersensitivity pneumonitis—An allergic disease of the lungs caused by inhaling various allergenic dusts.

hypersensitivity—A condition in which the immune system reacts to antigens that cause tissue damage and disease.

hyphae—A microscopic filament of cells that represents the basic unit of a fungus. They usually do not exist with yeasts.

immune—A condition in which an organism is protected against, or free from, the effects of allergy or infection, either by already having had the disease or by inoculation.

immunoglobulins—One of a family of proteins to which antibodies belong.

in vitro—In an artificial environment.

in vivo—In a live organism.

incidence—Number of new diseases occurring in a given population at a specified time period.

isotropic—Substances showing a single refractive index at a given temperature and wavelength, no matter what the direction of light may be.

itch mites—Aachnids with four pairs of legs which cause dermatitis and are sometimes implicated in house dust allergies.

ligase—Protein which acts as a glue to hold bits of DNA molecules together.

lipids Fats and fatty-like materials that are generally insoluble in water.

lymphocyte—A white blood cell important in immunity; of the two major types (both types have several subclasses), T lymphocytes are processed in the thymus and are involved in cell—mediated immunity, and B lymphocytes are derived from the bone marrow and are precursors of plasma cells, which produce antibody.

macrophage—A scavenger white blood cell that plays a role in destroying invading bacteria and other foreign material. It also plays a major role in the immune response by processing or handling antigens and as an effector cell in delayed hypersensitivity.

meningitis—An infection or inflammation of the membranes covering the brain and spinal cord. Onset symptoms are characterized by severe headache, stiffness of the neck, irritabilitiy, malaise, and restlessness, followed by nausea, vomiting, delirium, and disorientation. This progressed to increased temperature, pulse rate, and respiration. Nerve damage may culminated in deafness, blindness, paralysis, and/or mental retardation.

meningitis—Inflammation of any or all of the meninges (enclosing membranes) of the brain and spinal cord, usually caused by bacterial infections.

methylation—The process of combining with methyl alcohol.

***micrometer* (or micron)**—A unit of length equivalent to 10^{-6} meters.

***micron* (or micrometer)**—One-millionth of a meter (10^{-6}).

mold—Microscopic plants, belonging to the Fungi Kingdom, which do not have stems, roots, or leaves and are composed of a vegetative threadlike element (hyphae) and reproductive spores.

monoclonal antibody—Antibody produced by a single clone of cells that binds to only one epitope.

mushroom—Microscopic plants, belonging to the Fungi Kingdom, which are filamentous in nature with large fruiting bodies (referred to as the mushroom cap) which discharges reproductive spores.

mutagenic—Capable of causing a genetic change.

mycelium—A visible mass of tangled filaments of fungal cells, the rooting structure and extension of the more aerial hyphae.

mycotoxins—Toxins produced by molds.

nanometer—One one-thousandth of a micron, or one billionth of a meter (e.g., 10^{-9}).

nasal congestion—Blockage of the nasal passages.

necrosis—Death of living tissue.

necrosis—Localized tissue death.

neutralism—A lack of interaction between two microbial populations.

nonpolar compounds—Compounds which have atoms do not have a denser electron cloud about one atom, or group of atoms, than around its adjoining atom, or group of atoms (e.g., hexane).

nonviable—Not living, dead, destroyed organisms; not capable of causing disease.

nosocomial—Hospital acquired infections.

nucleotide—The sugar, phosphate, and base component of a DNA or RNA structure.

oligonucleotide—A short chain of specifically ordered nucleotide bases.

palynologist—A pollen specialist.

paper mites—Fictitious term with possible reference to storage mites or booklice.

parasitism—One population benefits and normally derives its nutritional requirements from the population that is harmed.

patch test—Used to identify substances responsible for contact allergy. The test consists of applying a small amount of a suspected substance to the skin. The area is covered with tape and left for forty—eight hours. If a small area where the substance was applied swells and turns red, the test result is said to be positive.

PCR—Acronym for polymerase chain reaction.

Petri plate (or dish)—A glass or plastic dish which has a lid and is used to isolate and grow microbes.

plasmids—Small loops of DNA in bacteria (contain genes for antibiotic resistance).

pleurisy—Inflammation of the pleura (or membranous sacs surrounding the lungs).

polar compounds—Compounds which have atoms which have a denser electron cloud about one atom, or group of atoms, than around its adjoining atom, or group of atoms (e.g., butanol).

pollen—Male fertilizing elements of a plant that is microscopic in size. Pollen grains are spheroid, ovoid, or ellipsoid in shape and may have a smooth, reticulated, spiculated, or sculptured surface.

polyclonal antibody Antibody produced by heterogeneous population of antibodies that bind to many different epitopes.

polymerase—An enzyme which aids in the linking of nucleotides to form DNA, RNA, or duplicated target sequences.

predation—One organism, the predator, engulfs and digests another organism, the prey.

predicted air concentrations—A calculated prediction of air concentrations of a given substance or substances, based on component emission rates.

prevalence—Normal frequency of a disease in a given population.

prick method—An allergy test whereby the skin is pricked with a needle at the point where a drop of allergen has been placed, introducing the allergen to the body's immune system.

primer—A synthesized short chain of predesigned, made-to-order genetic sequencing of deoxynucleotides bases which free in solution.

probe—A synthesized short chain of predesigned, made-to-order genetic sequencing of deoxynucleotides bases which is affixed to a substrate; same as a primer which is not affixed to a substrate.

pyrexia—A condition resulting in fever.

pyrocanic—Sensation of heat.

ragweed—A plant belonging to the family Compositae that is the major cause of hay fever in the United States. The ragweed family is large, with approximately 15,000 species.

refractive index—The ratio of the velocity of light in a vacuum to the velocity of light in a given medium. Higher atomic number generally results in a higher refractive index.

restriction endonucleases—Enzymes that "cut strands of DNA" at specific points (e.g., genetic scissors) used to isolate desired sections of genetic sequencing nucleotides.

restriction enzymes—Proteins that cut apart DNA molecules.

rheumatic fever—An inflammatory disease that usually occurs to young schoolage children and may affect the brain, heart, joints, skin, or subcutaneous tissues. Early on, it is characterized by fever, joint pains, nose bleeds, abdominal pain, and vomiting, progressing to chest pain, and, in advanced cased, heart failure.

rheumatoid arthritis—A chronic, destructive, often deforming, collagen disease which results in inflammation of the bursea, joints, ligaments, or muscles. It is characterized by pain, limited movement, and structural degeneration of single or multiple parts of the musculoskelatal system.

rhinitis—A disease of the nasal passages that is characterized by attacks of sneezing, increased nasal secretion, and stuffy nose (caused by swelling of the nasal mucosa).

RNA—Transfers genetic coded messages to provide instructions for cell maintenance and growth.

rust—A fungal disease of agricultural crops so named because of the orange-red color it imparts to infected plants. Belongs to the same fungal class as the common mushroom.

saponification—The hydrolysis of an ester by an alkali, producing a free alcohol and an acid salt.

saprophytic—Lives on and derives its nourishment from dead or decaying organic matter.

scratch test—A small drop of the allergen is applied over an area of the skin where a superficial scratch has been made. This allows the allergen to penetrate the top layer of skin.

semivolatile—Compounds with a boiling range of 240° C to 400° C.

sensitizee—To administer or expose to an antigen provoking an immune response so that, upon later exposure to that antigen, a more vigorous secondary response will occur. An individual can be immune (e.g., protected against the effects of an infectious agent or antigen) and sensitized to the antigen (e.g., demonstrate a positive tuberculin reaction) at the same time.

septicemia—A systemic, blood distributed disease which is caused by pathogenic organisms or their toxins.

septicemia—Systemic infection where pathogens are spread through the bloodstream, infecting various parts of the body, characterized by fever, chills, prostration, pain, headache, nausea, and/or diarrhea.

skin test—A method of testing for allergic antibodies. A test consists of introducing small amounts of the suspected substance, or allergen, into the skin and noting the development of a positive reaction (which consists of a wheal, swelling, or flare in the surrounding area of redness). The results are read fifteen to twenty minutes after application of the allergen.

slime mold—Organisms, belonging to the Fungi Kingdom, which are protozoan-like for part of their life cycle and reproduce by forming stalks which produce spores (or multiple spore-containing sporangia).

smut—A fungal disease of agricultural crops so named because of the sooty black appearance it imparts to infected plants. Belongs to the same fungal class as the common mushroom.

solid waste—Any garbage, refuse, sludge (from wastewater treatment plant, water supply treatment plant, or air pollution control facility), or other discarded material (including solid, liquid, semisolid, or contained gaseous material) resulting from industrial, commercial, mining, agricultural operations, and/or community activities. This definition excludes solid or dissolved material in domestic sewage, solid/dissolved materials in irrigation return flows or industrial discharges which are source/point source discharges subject to permits under Section 402 of the Federal Water Pollution Control Act, special nuclear/nuclear by-product material as defined by the Atomic Energy Act, Section 1004(27) of RCRA.

sorbent—A solid or liquid material which collects liquid and/or gaseous substances.

sporangiospores—Asexual produced spores which developed within a sporangium.

sporangium—A structure, or sac, within which spores develop.

spores—Reproductive cells of certain plants and organisms. Inhaled fungal spores are frequently the cause of allergic symptoms such as rhinitis and asthma.

storage mites—Four-legged arachnids which are sometimes implicated in outdoor environmental allergies associated with agricultural environments.

symbiotic—Obligatory relationship between two populations that benefits both populations.

synergistic—Mutually cohabitate with one another.

teleospores—Asexual spores produced by rusts.

thymine—A DNA base which only pairs to the base adenine.

tobacco sensitivity—Many people suffering from rhinitis and asthma experience heightened symptoms when exposed to tobacco smoke.

transparency—Ability to transmit light.

ulceration—Formation of a circular, crater-like lesion of the skin or mucous membranes.

unculturable—Viable and nonviable microorganisms which cannot be grown.

urediospores—Asexual spores produced by smuts.

urticaria (hives)—This is a skin rash characterized by areas of localized swelling, usually very itchy and red, and occurring in various parts of the body. It usually lasts only a few hours and involves only the superficial areas of the skin.

viable—Living organism; capable of causing disease.

viron—An intact virus particle which has the ability to infect.

virus—A submicroscopic organism which consists of genetic material and a coating and requires living organisms in order to reproduce

viscera—The internal organs enclosed within a body cavity, primarily the abdominal organs.

VOC—Volatile organic compound(s).

volatile—Compounds with a boiling range of 0° to 290° C.

yeasts—Species of fungi that grow as single cells.

zeophylic—Dry-loving organism.

Appendix 1

ABBREVIATIONS/ACRONYMS

AAAAI	American Academy of Allergy, Asthma, and Immunology
ACGIH	American Conference of Governmental Industrial Hygienists
AIDS	Auto Immune Deficiency Syndrome
ASHRAE	American Society of Heating, Refrigerating and Air-conditioning Engineers
BA	Blood Agar
CDC	Center for Disease Control
CFU	Colony Forming Units
CFR	Code of Federal Regulations
DOT	Department of Transportation
EDXRA	Energy Dispersive X-ray Analyzer
EMA	Electron Microprobe Analyzer
EPA	Environmental Protection Agency
EU	Endotoxin Units
FAME	Fatty Acid Methyl Esters
FDA	Food and Drug Administration
FID	Flame Ionization Detector
GC	Gas Chromatography
GC/MS	Gas Chromatography/Mass Spectroscopy
HEPA	High Efficiency Particulate Airfilter
HRP-SA	Horseradish Peroxidase-Streptavidin
IAQ	Indoor Air Quality
IMA	Ion Microprobe Analyzer
IP	Indoor Pollutant Methods (EPA)
KLARE	Kinetic-Turbidimetric Limulus Assay with Resistant-parallel-Line Estimates (Test)
LAL	Limulus Amebocyte Lysate
MIS	Microbial Identification System
NAB	National Allergy Bureau (Program under the American Academy of Allergy, Asthma, and Immunology)
NIOSH	National Institute for Occupational Safety and Health
OSHA	Occupational Safety and Health Act
PCR	Polymerase Chain Reaction
PID	Photoionization Detector

RBA	Rose Bengal Agar
RCRA	Resource Conservation and Recovery Act
SAED	Selected Area Electron Diffraction
SEM	Scanning Electron Microscopy
TCLP	Toxicity Characteristic Leaching Procedure
TEM	Transmission Electron Microscopy
TLV-TWA	Threshold Limit Value—Time-Weighted Average
TO Methods	Toxic Organic Methods
TSA	Tryptic Soy Agar

Appendix 2

UNITS OF MEASUREMENT

VOLUME

1 liter (l) = 1.06 quarts
1 milliliter (ml.) = 10^{-3} liter
1 microliter (μl) = 10^{-6} liter

LENGTH

1 meter (m) = 3.281 feet = 39.37 inches
1 centimeter (cm) = 10^{-2} meter = 0.039 inch
1 millimeter (mm) = 10^{-3} meter
1 micrometer (μm) = 1 micron (μ) = 10^{-6} meter

WEIGHT

1 gram(s) (gm) = 0.035 ounce
1 milligram (mg) = 10^{-3} gram
1 microgram (μg) = 10^{-6} gram

TEMPERATURE

1° Fahrenheit (F) = [(1.8) (X^{o} C)] + 32
1° Centigrade (C) = [X^{o} F — 32]/1.8

SAMPLE UNITS

ppm = parts of contaminant per million parts of sample material (e.g., air)
ppb = parts of contaminant per billion parts of sample material
mg/m^3 = milligrams of contaminant per cubic meter of sample material
$\mu g/m^3$ = micrograms of contaminant per cubic meter of sample material
grains/m^3 = grains of pollen per cubic meter of air sample

CFU/m^3 = colony forming units per cubic meter of air
mg/m^2-hour = milligrams of contaminant per square meter of material in one hour
mg/X-hour = milligrams of contaminant per item, or composite unit, (X) in
mg/hour-m^3 = milligrams of contaminant emitted in one hour within a three cubic meter space
μg/X-hour = micrograms of contaminant per item, or composite unit, (X) in one hour of emissions testing

FLOW RATES

1 cubic centimeter per minute (cc/min.)
1 liter per minute (lpm)

Appendix 3

POLLEN SAMPLING EQUIPMENT

The "relative cost" is a rated cost to allow for price fluctuations within the economy. The least expensive rating is less than 1, and the most expensive rating is greater than 5.

BURKHARD SPORE TRAP

Size: 14.5 inches (Recording Air Sampler) to 37 inches high (Seven-Day Recording Volumetric Spore Trap)
Weight: 20 pounds (Recording Air Sampler) to 32 pounds (Seven-Day Recording Volumetric Spore Trap)
Relative Cost: 1 to 4
Rent: Yes
Flow Rate: 10 lpm
Sample Duration: 24 hours (Recording Air Sampler only) or 7 days (Recording Air Sampler and Seven-Day Recording Volumetric Spore Trap)
Sample Media: Microscope slide coated with hexane/Vaseline.

Seven-Day Recording Volumetric Spore Trap (Courtesy of Burkhard Manufacturing Company Limited, Hertfordshire, WD3 1PJ, England)

Comments: Suction air sampler where pollen/spores are impacted on adhesive-coated transparent plastic tape, retained within a rotating drum. The Volumetric Spore Trap is not to be mistaken for the Portable Air Sampler for Agar Plates (a sieve impactor also sold by Burkhard Manufacturing Company). Evaluated and approved by the American Academy of Allergy, Asthma, and Immunology.

Pros: Refined, precision sampler. Highly regarded by researchers. Reliable and easy to use.

Cons: No known distributors in the United States.

KRAMER-COLLINS SPORE SAMPLER

Size: 8 inches in height
Weight: 3.5 pounds plus pump (3 to 16 pounds)
Relative Cost: less than 1
Rent: No
Flow Rate: 0.5 to 28.3 liters per minute
Sample Duration: 24 hours to 7 days
Sample Media: Double-coated cellophane tape backed on a slide.

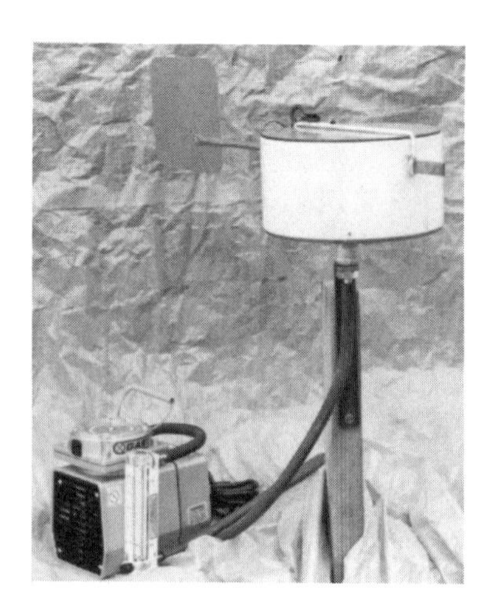

Kramer-Collins Seven-Day Spore Sampler
(Courtesy of The GR Electric Manufacturing Co., Manhattan, KS)

Comments: Similar to the Burkhard Air Sampler. Evaluated and approved by the American Academy of Allergy, Asthma, and Immunology.

Pros: Inexpensive, light weight, small, and marketed since 1976.

Cons: Distributed only by the manufacturer.

ROTOROD

Size: 3 inches by 4.75 inches by 3.5 inches (Model 20); 7 inches by 8 inches by 8 inches (Model 40)
Weight: 1 pounds (Model 20); 6 pounds (Model 40)
Relative Cost: 1 (Model 20) to 2 (Model 40)
Rent: Yes
Flow Rate: 2,400 rotations/minute (Type "I" rods: 120 lpm)
Sample Duration: American Academy of Allergy and Immunology recommends 24 hour interval sampling with a built-in timer (Model 40) or programmable, multi-control cycling field timer (separate purchase for the Model 20). Baseline intervals are 1 minutes on and 9 minutes off, and the sampling duration is adjusted according to anticipated loading.
Sample Media: Type "I" glass rod, coated with silicone grease.

Rotorod Model 20 with Cycling Field Timer and Retracting Sample Head
(Courtesy of Sampling Technologies, Inc., Minnetonka, MD)

Comments: Most commonly used equipment by allergists. The Type 'I' rods are coated with a thin coating of silicone grease to retain impacted pollen grains and mold spores as the rod rotates. When sampling is completed, the rods are removed from the head and mounted on a grooved stage adapter for microscope examination. For pollen identification, the exposed rods are treated with Calberla's stain to enhance grain details.
The portable Model 20 is used for short term, indoor and outdoor continuous air samples or interval sampling (when coupled with a cycling field timer). The larger Model 40 is used for long term, outdoor air samples, either with continuous use or at intervals. Both models may be fitted with a retractable head which is used for interval sampling or for sample completion protection to avoid impaction by wind/HVAC air movement when the timer discontinues sampling.

Pros: As the most commonly used equipment by local allergists and reporting entities, the results may be more comparable to those of local reports. The method is simple, equipment easy to use. Evaluated and approved by the American Academy of Allergy, Asthma, and Immunology.

Cons: As the rod moves air from its path, some of the pollen/spores are aerodynamically thought to be lost due to the air movement.

ALLERGENCO

Size: 6 1/8 inches by 3/4 inch by 5 inches
Weight: 4 pounds
Relative Cost: 1
Rent: No
Flow Rate: 15 lpm (built-in motor and flow control)
Sample Duration: 1 minutes to 24 hours (programmable); can be programmed to collect up to twenty-four discrete samples, eight per day, to allow for a three-day weekend sampling period.
Sample Media: Microscope slides covered with silicone grease.

Comments: This is an air suction device which draws air in and deposits airborne particles (e.g., pollen and spores) onto a microscope slide. Treated-microscope slides are placed on a solid brass holder within the unit, and the timer is set for the desired sampling time. When sampling is completed, the sample slide is removed and sample stained for analysis.

This unit can be programmed for continuous sampling or collection of up to 24 separate discrete samples deposited hourly at different sites for tracking trends. Sample results using the Allergenco tend to be three to five times the pollen/spore counts that of the Rotorod. Consequently, the results of this sampler should not be compared with those of other samplers.

Allergenco Air Sampler MK-3
(Courtesy of Allergneco-Blewstone Press, San Antonio, TX)

Pros: Easy to use.

Cons: Although used extensively, this unit is not specifically approved by the American Academy of Allergy, Asthma, and Immunology. Not comparable to the Rotorod, therefore there is generally no relevant means to compare results of reported counts by most allergists to that of the Allergenco Air Sampler.

Appendix 4

MICROBIAL SAMPLING EQUIPMENT

SIEVE IMPACTOR (i.e., Andersen Impactors)

Size: Portable
Weight: Variable (most of the weight is in the high volume sampling pump)
Relative Cost: 2 (single-stage); 2 (two-stage); 3 (six-stage)
Rent: Yes
Flow Rate: 28.3 lpm
Sample Duration: 60 seconds to 2 minutes (single- and two-stage); up to 5 minutes (six-stage)
Sample Media: Culture media in Petri dish(es).

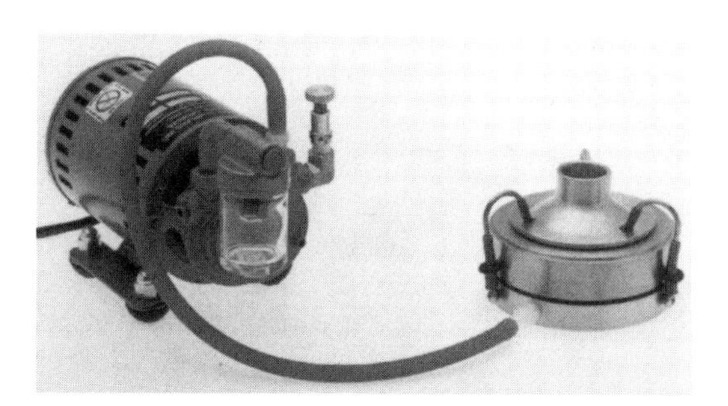

Single-Stage Andersen Impactor with Pump
(Courtesy of Graseby-Andersen, Atlanta, GA)

Comments: An air sampling pump capable of a flow rate in excess of one cubic foot per minute is calibrated to flow at 28.3 liters per minute (or the connector hose may be fitted with a pre-calibrated critical orifice). The unit(s) is (are) disassembled and fitted with a properly sized nutrient-containing Petri plate, face up with the lid removed.

A six-stage sampler requires six Petri plates. A two-stage sampler requires two Petri plates, and single-stage sampler takes one Petri plate. After installation, the unit is secured with retaining devices and the vacuum pump turned on for 1 to 5 minutes, depending of the unit used and anticipated loading.

In the six-stage impactor, the airborne particles are distributed according to particle size. In the two-stage impactor, the particle distributions are based upon respirable and nonrespirable sizes. In the single-stage impactor, all particles are collected indiscriminate of size.

Upon completion, the lids are replaced on the Petri plates and taped shut. If air shipment is anticipated, samples should be sent the same day they were collected, and arrangements for next day receipt by the analytical laboratory should also be made.

Pros: The Andersen impactors are the most commonly used samplers by experienced environmental professionals. Numerous studies have been performed to compare samplers, and the single-stage and six-stage Andersen impactor recovery results for "bacteria" are typically comparable to one another, along with the AGI-30 impingers.

In another study, using "mold spores," the six-stage Andersen impactor was found to be the most reliable of four sampler types in terms of precision and reproducibility of results. Other Andersen impactors were not studied.

Cons: Although there has been some research in this area indicating that the time of day is not as relevant as activities in progress at the time of the sampling, there is still some concern that the limited sampling duration does not provide the entire picture.

SLIT-TO-AGAR IMPACTOR (i.e., M/G Air Sampler)

Size: 12 inches by 10 inches by 12 inches
Weight: 16 pounds
Relative Cost: 2
Rent: No
Flow Rate: 28.3 lpm
Sample Duration: 5 to 60 minutes
Sample Media: Culture plate.

MG Air Sampler Model 220
(Courtesy of Barramundi Corporation, Homosassa Springs, FL)

Comments: The slit-to-agar sampler is also referred to in literature as the SAS or STA. Air is drawn through a slit at a rate of one cubic foot per minute and impinged upon an agar surface just below the slit.

There are two models (e.g., Model 220 and P-320), and both have interchangeable motor drives. The Model 220 sampler may be rotated at one of four drive speeds—one revolution per 5, 15, 30, or 60 minutes.

Pros: Limitation of area impacted allows for sample durations up to one hour.

Cons: In one sampler comparative study, bacteria recovery for the Mattson-Garvin SAS was poor. In another sampler comparative study, fungal spore recovery for the SAS was irratic, dependent upon concentration levels. Although consistently lower recoveries occurred, the SAS appears to have been within the same range as the greatest collection efficiency, and improvements have been made by the manufacturer since the last study was performed.

CENTRIFUGAL IMPACTOR (i.e., RCS Plus)

Size: Portable
Weight: less than 4 pounds
Relative Cost: 3
Rent: Yes
Flow Rate: 50 lpm
Sample Duration: Time required to collect 1 to 1,000 liters of air at 50 lpm
Sample Media: Agar media strips.

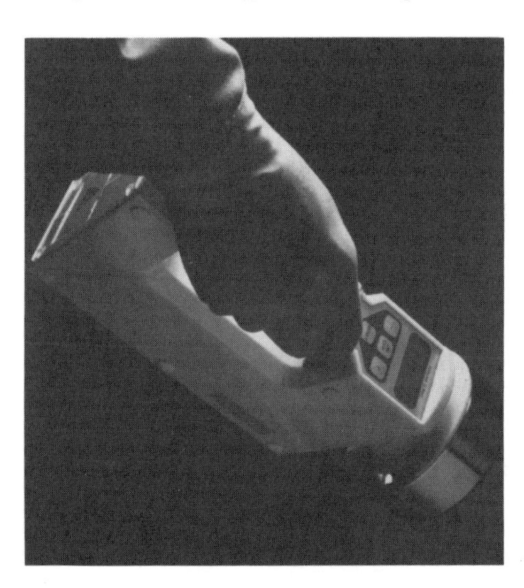

RCS Plus Centrifugal Air Sampler (Courtesy of Biotest Diagnostics Corp., Denville, NJ)

Comments: The centrifugal impactor is sometimes referred to in literature as a Reuter Centrifugal Sampler, or RCS. Select the desired air volume by depressing the + and – buttons on the display pad. Remove an agar strip from its protective wrapper, and insert it in the rotor assembly. Turn it on by pressing the on/off button. When the unit automatically turns off, remove the agar strip and return it to its protective sleeve. In the Standard RCS, the air enters and leaves through the top of the impeller drum. In the RCS Plus, the air enters the drum at one port and exits another.

The user's manual for the Standard RCS provides a formula for determining the results, based upon the sampling time. The RCS Plus is a more direct measurement and calculation. For other differences, consult the manufacturer.

Pros: Easy to use. The RCS is battery operated, light weight, and easily carried by hand. Thus, it may be used to probe areas which might otherwise not be accessible.

Cons: Requires special agar strips, provided only by the manufacturer of the instrument.

LOW VOLUME FILTRATION

Size: Portable
Weight: Personal sampling pump (around 3 pounds)
Relative Cost: 1
Rent: Yes (personal sampling pumps)
Flow Rate: 1 to 2 lpm
Sample Duration: up to 8 hours
Sample Media: Polycarbonate filter generally preferred.

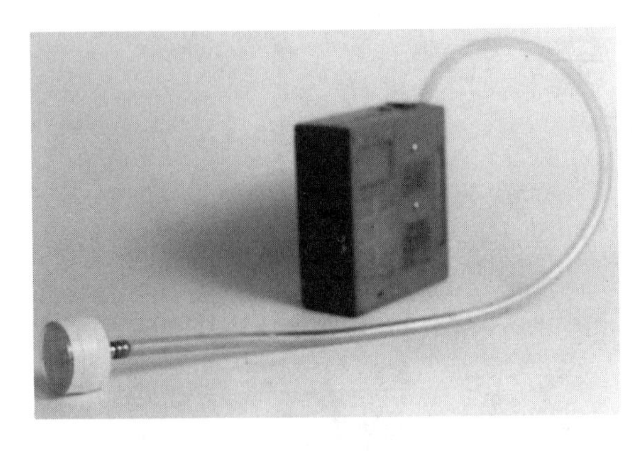

Low Volume Air Sampling Pump with Sampling Filter

Comments: Filter collected dust/particles are transferred to a Petri dish through direct transfer of a filter wash solution onto the nutrient agar or retention of the wash solution for multiple plating. Multiple plating capabilities allows for dilutions of otherwise excessive microbial numbers and for the potential plating of the same sample onto several different nutrient agar plates.

Pros: Easy, inexpensive, greater sample duration (more representative of an entire day).

Cons: Probable loss of microbial viability due to drying effect and impaction onto the filter. In a previously mentioned study, filtration of two different types of bacteria had different results. The bacteria *Escherichia coli* was not recovered, but the bacteria *Bacillus subtilis* recovery was equivalent to the recovery by the Andersen impactor. Although bacterial sampling is questionable, filtration sampling is more feasible for protected spores than the more fragile bacteria types.

HIGH VOLUME AIR FILTRATION
(i.e., Andersen PM-10)

Size: 21 inches in diameter by 61 inches high
Weight: Around 100 pounds
Relative Cost: >5
Rent: Yes
Flow Rate: 1,136 lpm (ranges from 140 to 1,400 lpm)
Sample Duration: Variable (up to 24 hours); programmable timer can be set to sample intermittently for any given time period.
Sample Media: Special high volume filter.

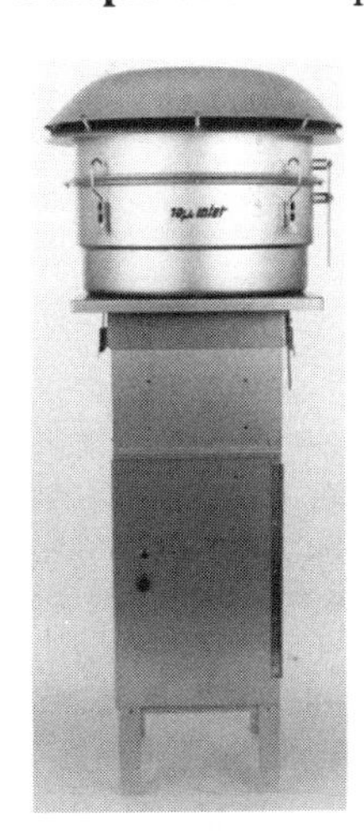

Andersen Size Selective Hi Vol Series PM-10
(Courtesy of Graseby-Andersen, Atlanta, GA)

Comments: This sampler is typically an outdoor sampler. It should be set up and operated according to the manufacturer's instructions.

Collected dust/particles are transferred to a Petri dish through a wash transfer. The wash solution can also be used for multiple samples, and one sample may be plated onto several nutrient agar plates in the same manner that the filter samples are manipulated where low volume filter sampling has been performed.

Pros: Collects extremely high air volumes.

Cons: Considerable loss of microbial viability due to drying effect and impaction onto the filter. The velocity alone may prove damaging to even the hardiest of microbes. Difficult to transport and relatively expensive unless the equipment is available, used for other purposes.

IMPINGERS (i.e., AGI-30)

Size: Portable
Weight: Variable (dependent on high volume pump)
Relative Cost: 1
Rent: No
Flow Rate: 6 or 12.5 lpm (dependent upon stem length)
Sample Duration: 30 minutes
Sample Media: Sterile dispersant (or surfactant) solution.

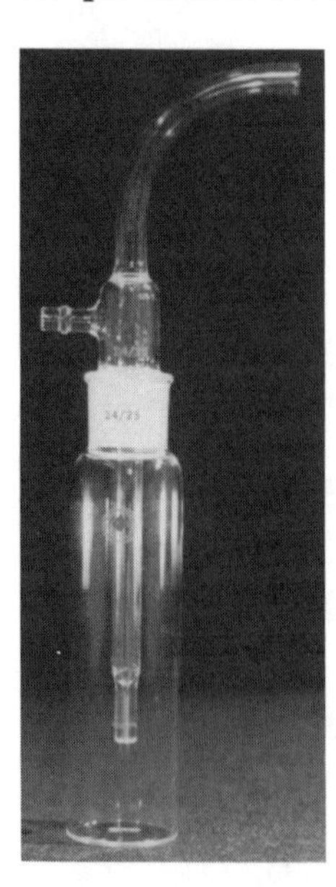

AGI-30 Impinger
(Courtesy of Ace Glass, Vineland, NJ)

Comments: The AGI-30 is a much larger impinger than that which is normally used in industrial hygiene sampling and is designed to manage higher flow rates, without sample solution losses. Yet, some feel that the smaller impinger is comparable at a 2.0 ± 0.5 lpm. Whichever impinger is used, however, it will require sterilization which is generally performed by the laboratory (and most labs provide their own sterilized AGI-30 impingers).

Place 20 milliliters of sterile dispersant solution into a previously cleaned, sterile impinger. With tubing, connect the side arm of the impinger to a high volume sampling pump. Sample for 30 minutes at a preset flow rate as designated by the manufacturer (either 6 or 12.5 lpm). Upon completion of sampling, transfer the sample solution to a sterile container. Seal the container and send for analysis. If air shipment is required, send it overnight.

Laboratory handling procedures vary. Frequently published methods have involved aspiration of the solution through a filter and either treatment with nutrient media or placement of the filter on nutrient contained within a Petri dish.

Other processing methods have been suggested and seem to follow a more solid line of logic. The solution provides sufficient material to be plated for taking several aliquots of sample and plating them on different nutrient agar plates. Then, too, should there be an excess of material, the sample can easily be diluted prior to plating.

Pros: Easy, inexpensive. Results consistently come close to the impactor samples in comparison, efficiency sampling. The results are typically just short of the six-stage impactor, but slightly higher than the single- and two-stage samples for bacteria.

Cons: Excessive shear force involved in collection as the airstream approaches sonic velocity which tends to cause the destruction of some vegetative cells.

The "relative cost" is a rated cost to allow for price fluctuations within the economy. The least expensive rating is less than 1, and the most expensive rating is greater than 5.

Appendix 5

ALLERGENIC DUST SAMPLING EQUIPMENT

STANDARD VACUUM CLEANER

Size: Variable
Weight: Variable
Relative Cost: 1
Rent: Yes
Flow Rate: Variable
Sample Duration: Sufficient to collect a minimum of 1/2 cup of dust.
Sampling Supplies: Vacuum bag, household vacuum cleaner, and secondary sample retention bag/container for shipping.

Comments: The final results have been comparable when comparing a standard vacuum versus a HEPA. The use of a HEPA or high efficiency bag, however, may provide additional capture competence without compromising simplicity. Use a different bag for each sample collection.

Composite sampling will help identify the probable allergen and discrete sampling will isolate the reservoir. For each discrete sample, either select a 9 cubic meter area (any clearly delineated area) or select a functional area (e.g., traffic areas on the carpeting). Remember that a minimum sample size is 200 milligrams (ideally 500 milligrams) of collected material.

Label and send each separate bag to the laboratory for analysis. Placement of each sample into a separate sealed, plastic baggy or other form of sealable container is strongly recommended.

Pros: Easy, inexpensive, fairly reliable (dependent upon sample sites chosen).

Cons: Discrete samples involve a shot-in-the-dark as to the area or areas where allergenic dust is generated. If there is a wrong guess, the results may easily be misinterpreted as nonproblematic. The analytical process is also a limitation in that if the specific allergenic dust is not properly identified (e.g., booklice for which there are no published assays), it may be bypassed and the results found negative for animal allergenic dust.

HEPA VACUUM CLEANER

Size: Variable
Weight: Variable
Relative Cost: 1
Rent: Yes
Sample Duration: Sufficient to collect a minimum of 1/2 cup of dust.
Sample Supplies: Vacuum bag, vacuum cleaner, and secondary sample retention bag/container for shipping.

Comments: The use of a HEPA assures capture of particles less than 1 micron in size. Be certain to use a separate collection bag of each sample. For sampling information, see the comments section under the heading of Standard Vacuum Cleaner.

Pros: Easy, inexpensive, fairly reliable (dependent upon sample sites chosen), greater collection efficiency.

Cons: This is a shot-in-the-dark as to the area or areas where allergenic problems are generated. If there is a wrong guess, the results may be inconclusive or misinterpreted as nonproblematic. In the latter case, the investigator goes on to the next possibility, assuming the proper location for sampling was chosen.

VACUUM ATTACHMENTS

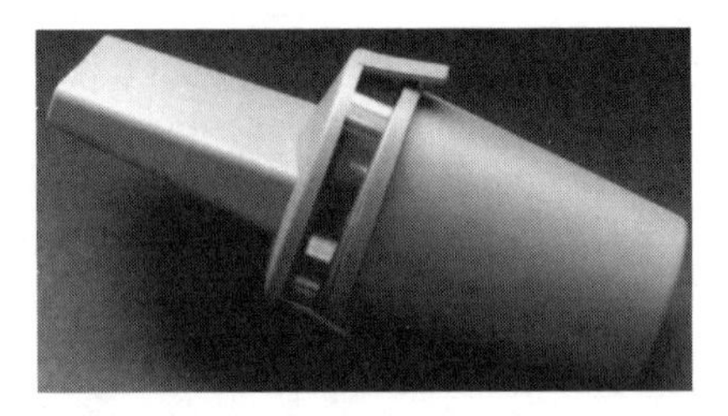

ALK Device
(Courtesy of Vespa Laboratories, an
ALK-Abello Company, Spring Mills, PA)

Comments: There are various vacuum attachments designed to reduce the potential for cross-contamination by contaminated hoses and other internal parts of the vacuum. A specially designed sample collection bag is relatively simple, inexpensive. The ALK device is a plastic wand with a built-in filtra tion/collection container. Neither the collection bag nor the wand collection container is reusable.

Pros: Minimize cross-contamination of samples. The sample collection bag is inexpensive (oftentimes supplied by the laboratory at an additional fee).

Cons: Requires prior purchase and planning. This is not something one does in a spur-of-the-moment response.

POLYCARBONATE MEMBRANE FILTER AND VACUUM PUMP

Size: Variable
Weight: Dependent upon the pump
Relative Cost: 1
Rent: Yes (air sampling pump)
Flow Rate: Irrelevant
Sample Duration: Variable
Sample Media: Polycarbonate membrane filter with a personal or high volume sampling pump.

Low Volume Air Sampling Pump with Sampling Filter

Comments: A personal or high-volume pump may be used to draw air and its associated dust through a filter. This method serves as a miniature, controlled vacuum cleaner. However, air volume is not relevant. The area and amount collected must be similar to that of the preceding methods.

Pros: Easy, inexpensive, contaminant-free collection.

Cons: The amount of dust present may require sampling over a large area which would unfeasible for a small collection sampler.

The "relative cost" is a rated cost to allow for price fluctuations within the economy. The least expensive rating is less than 1, and the most expensive rating is greater than 5.

Appendix 6

FORENSIC DUST SAMPLING EQUIPMENT

LIFTING SETTLED DUST

Sampling Supplies I: "Clear" tape, microscope slide, stain (e.g., Calberla's solution), and indelible marking pen.
Sampling Supplies II: Specialty sample collection tape.

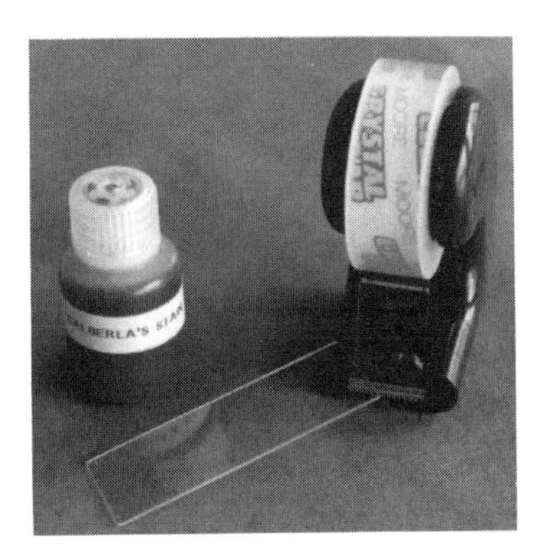

"Clear" Tape, Microscope Slide, and Calberla's Stain

Comments: The "clear tape" method involves dispensing the tape and touching the sticky surface to the dust. Keep in mind that a dust sample may be excessive. Look at the collected material to assure a light even dispersion of dust over the surface of the tape. If the sample is to be analyzed "as is," a drop of the stain should be placed on the slide prior to securing the sample. Tape the sample to the surface of the slide, and use an indelible pen to identify the sample on the slide.

The "specialty tape" method involves dispensing the tape and touching the sticky surface to the dust. Loading is not an issue where the sample is to be extracted from the surface of the tape by the analyst. After the sample has been lifted, the tape may be folded back on itself, labeled, and forwarded for analysis as is.

Pros: Inexpensive, easy.

Cons: Not easy to pick and choose particles from the "clear" tape. Limited area coverage.

MICROVACUUMING

Sampling Supplies: Polycarbonate, glass fiber, or mixed-cellulose ester membrane filter with either a low volume or high volume air sampling pump. If air sampling is to be performed, obtain pre-weighed filters.

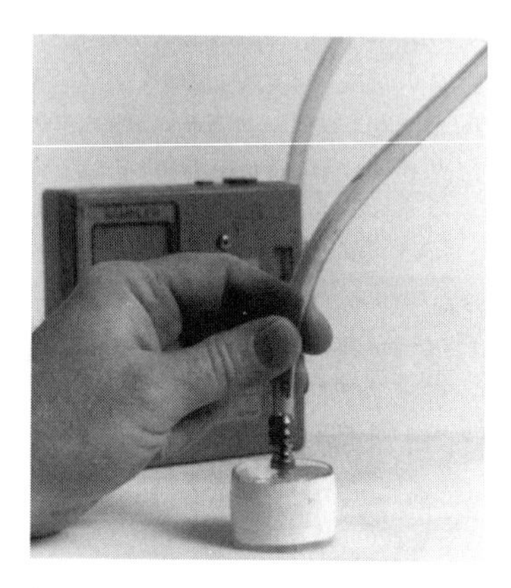

Low Volume Air Sampling Pump Used as a Vacuum with Sampling Filter

Comments: A sampling pump may be used to draw air and its associated dust through a filter. This method serves as a miniature, controlled vacuum cleaner. However, in dust collection air volume is not relevant. The area and amount collected must be similar to that of the preceding methods.

Where air sampling is to be performed, use a pre-weighed filter. Set a flow rate, note the start/stop times, and calculate the air volume. Although the major concern is for the total dust collected, a determination of air volume will provide extra information which may be usable for comparative samples. Samples may be lifted from the filter(s) at the laboratory for particle identification.

After the sample has been collected, seal, cap, and/or replace the cover to the ends of the filter. Label the filter, and send it to the laboratory for analysis.

Pros: Easy, inexpensive, contaminant-free collection. Controllable sample size. Potential for air sample collection as well as settled dust collection.

Cons: Requires equipment not readily available in the supermarket.

CYCLONE DUST SAMPLER

Sampling Supplies: Pre-weighed filter, cyclone, and air sampling pump.

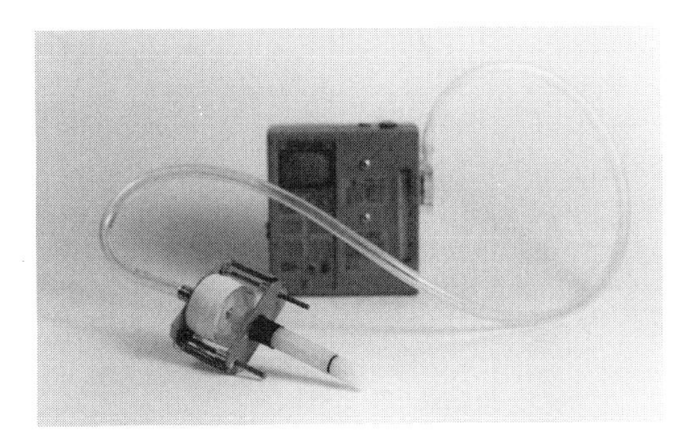

Cyclone Respirable Dust Sampler

Comments: This method may be used only for air sample collection. It allows for collection of respirable dust only.

Set the flow rate (using a containerized calibration technique with the cyclone connected by tubing to the sampling pump) at 1.7 lpm. Set the flow rate, note the start/stop times, and calculate the air volume.

Pros: As it provides a means for collecting respirable dust air samples, this method is more targeted to that which is respirable and airborne. Yet, the levels will require comparison with nonproblem areas as those limits which have been recommended within this book are for settled dust, not airborne dust.

Cons: This method is more time consuming as far as sample duration is concerned, and there has been minimal research performed regarding "airborne allergenic dust levels."

The "relative cost" is a rated cost to allow for price fluctuations within the economy. The least expensive rating is less than 1, and the most expensive rating is greater than 5.

Appendix 7

AMBIENT AIR SAMPLING EQUIPMENT

EVACUATED CANISTERS (i.e., Summa™ Canister)

Size: Variable (most are portable)
Weight: Variable
Relative Cost: 1 (canister only); 2 (canister with pump); >5 (flow control adapter)
Rent: Yes
Flow Rate: Depends on adapters for flow control.
Sample Duration: Instantaneous to 24 hours.
Sampling Supplies: Pre-cleaned, nonreactive evacuated canister with an open-close valve, controlled flow rate valve, or vacuum pump.

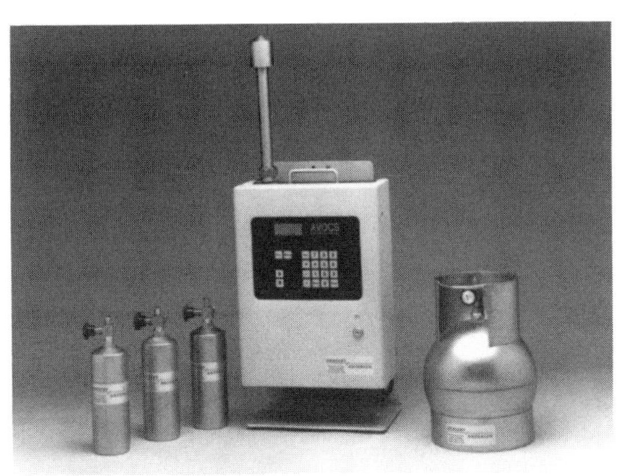

Summa™ Canister
(Courtesy of Graseby Andersen, Atlanta, GA)

Comments: This approach set the pre-cleaned canister in the area where sampling is to be performed and opens the valve. The valves allow for an immediate "grab" sample, or they may be set to allow for a specified flow rate and sample duration. If 24-sampling is anticipated, a 12-liter canister should be used. Allow the predetermined time to pass and close the valve. Send the canister to the analytical laboratory in a protective container to avoid damage. If feasible, send with a cold storage container (e.g., plastic cooler).

Pros: Easy to use and ship by air freight. Rapid sampling is possible while longer sample durations are an option as well.

Cryogenic processing permits concentration of air samples. The sample may represent as much as 1 liter of air. Large canisters permit several analytical runs. For example, a 6-liter Summa canister will provide up to six 1-liter samples. The thermal desorption sample gets one chance per sorbent tube.

Cons: Cryogenic processing tends to filter the polar compounds which are attracted to the water molecules that are filtered out prior to instrument processing.

BAG SAMPLER (i.e., Integrated Bag Sampler)

Size: Variable (10 to 120 liters)
Weight: Dependent upon retaining vessel
Relative Cost: 1
Rent: Doubtful
Flow Rate: Variable (0.2 to 20 lpm)
Sample Duration: Variable (1 minutes to 8 hours)
Sampling Supplies: Nonreactive sampling bag (e.g., Tedlar), retaining vessel, and personal or high-volume sampling pump.

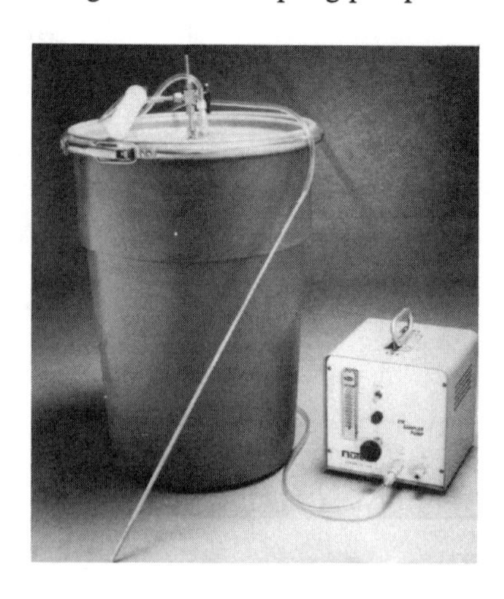

100-Liter Integrated Bag Sampler
(Courtesy of Graseby-Nutech, Durham, NC)

Comments: An integrated sampling vessel is recommended to avoid drawing air samples through contaminated pump systems. These vessels may be constructed of a nonflexible container and Teflon fittings or purchased from a manufacturer of specialty equipment.

The bag is placed in the vessel with an access port to the exterior. The vessel is closed, and a vacuum is created within, drawing the bag outward, allowing the exterior air to displace the void caused within the bag. When sampling is completed, all valves are closed. There are bags as large as 120 liters. In the higher volume samples, a retaining vessel the size of a plastic garbage can may be required which will take some resourcefulness if the environmental professional wants to collect the entire volume.

Pros: Easy, inexpensive.

Cons: Potential loss of sample through bag diffusion and leaks. If the bag is completely filled, shipping by air or under hot conditions may result in expansion of the bag contents and damage to the seams. There are also concerns that some chemicals adhere to the interior surface of the bag.

AMBIENT AIR SAMPLER

Size: 200 milliliter
Weight: Several ounces
Relative Cost: Provided at no additional cost by the analytical laboratory.
Rent: Supplied by lab
Flow Rate: Not applicable
Sample Duration: Instantaneous
Sampling Supplies: Evacuated ambient air sampler/test tube and squeeze bulb with special prong adapters.

Ambient Indoor Air Sampler
(Courtesy of Texas Research Institute, Austin, TX)

Comments: The ambient air sampler provides a low volume sample which may be used for screening purposes only. It involves inserting the retrofitted bulb prongs into the evacuated sampler and squeezing the bulb for a specified number of times (typically three times), withdrawing the tube, capping the sample, and returning the tube along with the sampler to the analytical laboratory.

Pros: Easy, inexpensive.

Cons: Limited in detection ability (down to 0.5 ppm, at best) and can only be used for screening.

The "relative cost" is a rated cost to allow for price fluctuations within the economy. The least expensive rating is less than 1, and the most expensive rating is greater than 5.

GENERAL INDEX

Bold faced page entries indicate associated figures, diagrams, tables, and some of the more important table components.

A

B

C

F

G

H

M

N

O

P

Q

R

T

U

V

W

X

Y

Z

SYMPTOM INDEX

A

B

C

D

E

F

P

R

S

T

U

V

W